Lecture Notes in Energy

Volume 36

Lecture Notes in Energy (LNE) is a series that reports on new developments in the study of energy: from science and engineering to the analysis of energy policy. The series' scope includes but is not limited to, renewable and green energy, nuclear, fossil fuels and carbon capture, energy systems, energy storage and harvesting, batteries and fuel cells, power systems, energy efficiency, energy in buildings, energy policy, as well as energy-related topics in economics, management and transportation. Books published in LNE are original and timely and bridge between advanced textbooks and the forefront of research. Readers of LNE include postgraduate students and non-specialist researchers wishing to gain an accessible introduction to a field of research as well as professionals and researchers with a need for an up-to-date reference book on a well-defined topic. The series publishes single and multi-authored volumes as well as advanced textbooks.

More information about this series at http://www.springer.com/series/8874

Charles A.S. Hall

Energy Return on Investment

A Unifying Principle for Biology, Economics, and Sustainability

 Springer

Charles A.S. Hall
College of Environmental Science
 and Forestry
State University of New York
Syracuse, NY
USA

ISSN 2195-1284   ISSN 2195-1292 (electronic)
Lecture Notes in Energy
ISBN 978-3-319-47820-3   ISBN 978-3-319-47821-0 (eBook)
DOI 10.1007/978-3-319-47821-0

Library of Congress Control Number: 2016954695

© The Author(s) 2017
This work is subject to copyright. All rights are reserved by the Publisher, whether the whole or part of the material is concerned, specifically the rights of translation, reprinting, reuse of illustrations, recitation, broadcasting, reproduction on microfilms or in any other physical way, and transmission or information storage and retrieval, electronic adaptation, computer software, or by similar or dissimilar methodology now known or hereafter developed.
The use of general descriptive names, registered names, trademarks, service marks, etc. in this publication does not imply, even in the absence of a specific statement, that such names are exempt from the relevant protective laws and regulations and therefore free for general use.
The publisher, the authors and the editors are safe to assume that the advice and information in this book are believed to be true and accurate at the date of publication. Neither the publisher nor the authors or the editors give a warranty, express or implied, with respect to the material contained herein or for any errors or omissions that may have been made.

Printed on acid-free paper

This Springer imprint is published by Springer Nature
The registered company is Springer International Publishing AG
The registered company address is: Gewerbestrasse 11, 6330 Cham, Switzerland

Preface

Most earlier civilizations, and many today, attributed their existence, well-being and rules for proper behavior to a powerful God or gods. Our civilization, and indeed earlier ones, has a new god, energy, rarely acknowledged but increasingly understood by a relative few, that allows our existence and determines many aspects of our material well-being. Energy is the master determinant of most that happens on Earth. As such it has determined, and will continue to determine, directly and indirectly, the major events of civilization. But this view is not well understood, principally because most analysts, economists and prognosticators, not to mention college freshmen, have not been taught the fundamentals of energy, which is too often presented, if at all, in a dispersed highly technical way that cannot reach most of the potential readers. I have spent my adult life studying energy and systems science, first as an ecologist and subsequently as something like an economist. I was trained by many but especially my Ph.D. advisor, Howard Odum, who opened a world of understanding and explanation to me that allowed me to understand nature as much more systems and process rather than the disparate hodgepodge of differing activities and species that I had been previously trained to see. This book is meant as a straightforward and relatively non-technical introduction to energy and its role in nature and in human-dominated systems, including economics. It is a primer on how the world works, emphasizing commonalities in structures and processes obvious to one trained in systems science but to relatively few others. My modest claim to scientific fame has come from the development of the concept Energy Return On Investment (EROI, sometimes EROEI) which I first elaborated while studying why fish migrate but later applied to many things, including especially our search for new energy. My goal is to present the needed information as a story in our understanding of life, evolution and human society in relatively few pages.

I found it surprising that others did not see things from this energy perspective, as it seemed to me so clear, so general and so powerful in explanation. When later I worked in some thirty different countries around the world, I was often asked how, after a day or two of introduction, I could bring such useful insight to the problems of that country I was visiting? I would respond: "But I have seen it before, many times, often with different actors but common processes, all with energy as a basis." For energy is ubiquitous, even if our understanding of it is not. Energy is not something that can be observed only in the physics lab or with expensive instruments but is part of everything we do. Take the kitchen: we may think of energy when we boil water or turn on the oven, but who thinks of a sharp knife as energy, as a force concentrating device, of cooking as a means of giving the human gut access to plants whose cell walls made most plants inaccessible to humans before the human use of fire, or the sunlight and food chains that made the food available in the first place? Nor do we usually think of the energy investments made by farmers and the entire industrial system to get the food to us in convenient and (relative to historical conditions) inexpensive form at the local supermarket.

This story focuses on energy investments and the return on those investments as a kind of underlying super process that guides essentially all that life, including humans, do. This is a story about what we have learned, and what we need to know, to understand our place in the universe and the future of human civilization.

Syracuse, USA   Charles A.S. Hall

Reference

Hall, C.A.S. 1972. Migration and metabolism in a temperate stream ecosystem. Ecology 53(4): 585–604. (Ph.D. Thesis, University of North Carolina, Chapel Hill, 1970).

Acknowledgements

Nearly all of my life's professional work has been with others, who generally are listed with me in my publications. I thank them again here. With respect to this book Jim Brown, John Day, Rod McNeil, Garvin Boyle, Ed Brothers, Ajay Gupta, Don Henderson, Solis Norton and Jacques Treiner reviewed sections, and François-Xavier Chevallerau, Britt-Marie Lindstrom, and Richard Vodra reviewed the entire book. Errors that remain are my fault. As always, I wish to acknowledge the very large influence of my great teacher, Howard Odum, the greatest mind I have ever had the privilege to know, and a real gentleman.

Contents

Part I Energy and Investments

1	**Investments**	3
	Reference	6
2	**The Development of the Laws of Thermodynamics**	7
	2.1 The History of Our Understanding of Energy	7
	2.2 Heat Energy	9
	2.3 Developing the Laws of Thermodynamics	10
	2.4 Understanding the Laws of Thermodynamics Today	14
	2.5 Implications of the Second Law	15
	2.6 The Second Law and Efficiencies	18
	Reference	20
3	**About Energy**	21
	3.1 What Is Energy?	23
	3.2 Biology	25
	3.3 Quantity of Energy	25
	3.4 Exergy and Emergy	26
	3.5 Quality of Energy	26
	3.6 Types of Energy	27
	3.7 Energy Density	28
	3.8 Power	29
	References	31
4	**Energy and the Physical World**	33
	4.1 Main Sources of Energy on Earth	33
	4.2 Hadley Cells	35
	4.3 "Investments" by Nature	38
	4.4 The Formation of Fossil Fuels	39
	4.5 Oil	41
	4.6 Natural Gas	42

	4.7	Unconventional Petroleum: The Fracking Revolution	42
	4.8	Coal	45
	References		46

Part II Energy and Biology

5 The Ecological Theater and the Evolutionary Play 49
    5.1 Energy and Biology 50
    5.2 Fuels 53
    5.3 Metabolism 55
    References 57

6 Energy Return on Investment as Master Driver of Evolution 59
    6.1 Darwinian Evolution 59
    6.2 Fitness 62
    6.3 What Determines the Distribution and Abundance
         of Species? 63
    6.4 Energy as the Master Resource for Evolution 65
    6.5 Energy Return on Investment as Master Driver 66
         6.5.1 A Special Section for Fly Fishers 67
         6.5.2 EROI and the Growth of Tits 70
    6.6 Energy Return on Investment as the Means of Obtaining
         Darwinian Fitness 71
    References 72

7 Maximum Power and Biology 73
    7.1 History 73
    7.2 Maximum Power for One Process 75
    7.3 Maximum Power for Systems 81
    7.4 Proposals for Maximum Power Principle
         as the 4th Thermodynamic Law 85
    7.5 Conclusion 85
    References 86

Part III Energy and Human Economies

8 Energy in Early Human Economies 89
    8.1 Application to Our Early Ancestors 89
    8.2 Hunter-Gatherers 90
    8.3 Agriculture and Deforestation 91
    8.4 Were Early Human Societies Sustainable? 92
    8.5 Expense of Energy to Early Civilizations 92
    8.6 EROI Analyses Over Very Long Periods of Time 93
    References 94

9	**Fossil Fuels**	95
	9.1 Economic Implications of Fossil Fuels	98
	9.2 Efficiency in Energy Use	99
	9.3 Fossil Energy and Economics	100
	9.4 The Role of BioPhysical Economics	101
	References	105
10	**EROI and Industrial Economies**	107
	10.1 Introduction	107
	10.2 Why Should EROI Change Over Time: Technology Versus Depletion	109
	10.3 What We Know About EROI Values and Trends for Different Fuels	111
	References	116
11	**Methods and Critiques for EROI Applied to Modern Fuels**	119
	11.1 How We Do EROI Analysis: More Detail	119
	11.2 Energy Return Data	122
	11.3 Energy Investment Data	122
	11.3.1 National Energy Accounts of Direct Energy Used	123
	11.3.2 National-Level Accounts for Capital Expenditures and Other Indirect Uses	124
	11.3.3 Process Analysis	125
	11.4 Quality Corrections	126
	11.5 Estimating EROI at Point of Use	127
	11.6 EROI of Obtaining Energy Through Trade	128
	11.7 Methodological Issues, Problems and Criticisms of EROI	129
	11.7.1 Criticisms of EROI Studies: Definitions	130
	11.7.2 Differences in Results: Corn-Based Ethanol	131
	11.7.3 Differences in Results: Photovoltaic Systems	132
	11.7.4 Corrections for Energy Quality	133
	11.7.5 Theoretical Versus Empirical Assessments of Electricity Output	134
	11.7.6 Boundaries and Comprehensiveness of the Cost Assessments	135
	11.7.7 Technological Changes Over Time	136
	11.8 Other Issues That Need Consideration That Might Decrease the EROI of Solar Energy: Storage	136
	11.9 Exponential Growth of Energy Production	137
	11.10 Summary of Critiques	138
	11.11 Further Issues in Comprehensiveness of EROI Analysis	138
	11.12 Business Services and Taxes	139
	11.13 Labor	139
	References	140

12	The History, Future, and Implications of EROI for Society	145
12.1	Sustainability	145
12.2	Peak Oil: How Long Can We Depend on Oil and Other Fossil Fuels?	146
12.3	New Technologies to the Rescue?	150
12.4	EROI	152
12.5	What Level EROI Does Society Need?	154
12.6	Economic Impacts of Peak Oil and Decreasing EROI	155
	12.6.1 Secular Stagnation	159
12.7	Developing Energy Policy	162
References		167
Index		171

Part I
Energy and Investments

Economists usually consider that there are two things we do with the resources at our disposal, consumption (using them now, often for our own pleasure) or investments (defering their use and using them in some way to generate more resources in the future). They, and people more generally, usually think of such resources and investments in terms of money or time. But in essentially all cases investments must be made in terms of energy. This section introduces the concepts of resources and energy, starting with an historical approach.

Chapter 1
Investments

Our story starts with investments. The concept of investment, of using resources already in hand to attempt to get more in the future, is familiar and essential to the human condition, indeed to all of life. It can explain a tremendous amount of our behavior. At the level of ourselves, our families, our communities, our nation or our world everything, in a sense, is about investments and return on those investments. The resources required for the generation of our food, for such wealth as we are able to garner, indeed for life itself do not appear for free but usually must be sought through investments, large or small. This is most obvious as we grow from children to young adults and observe the adult roles we might take on for ourselves. Farmers, bankers, entrepreneurs, most business people, musicians, and even academics are well aware that if they are to do well in their chosen field very large investments in time, effort, money, study, practice, and so on are necessary. The concept of return on investments is also very familiar to us as we examine financial returns on stocks or savings or houses or family. Each partner in a relationship must invest in various ways time, energy, thoughtfulness, and love or at least care if the relationship is to be successful. Various professional or charitable institutions emphasize the importance of investing in children, good government, science, or many other things for society's well-being. The list is almost endless because the concept of investment is crucial for nearly all that we do. Clearly, investments and return on investment underlies most of what we do and can do in life. Most of us know people who are obsessed with trying to get some kind of high return on their own investment. An investment implies something to be invested, which necessarily must be a surplus from the past. This book is about how we can generalize on the concept of return on investment and how we can understand nearly all investments much better by understanding them in terms of energy.

Nothing happens without energy. This applies to economic activity and investments just as truly as to moving a soccer ball. But energy does not come for free, for many resources, including energy, must be invested to get energy for a later date. Gardeners know this: one has to till and fertilize the soil, plant the seed, water the little plant carefully and wait patiently for the plant to mature, usually while investing in various means to protect your investment from other species (or your own) eager to exploit your own work. Natural history, and human history, usually can be viewed as various species or groups of people making investments in an attempt to get more resources. Much of the human history of the world is about making investments, of necessity energy investments, in resource exploitation, trade and shipping, military campaigns, and so on in an attempt to secure more. But, importantly, throughout most of history, indeed for most of human civilization, energy was expensive, as it is expensive too for other organisms. A recent paper by King et al. (2015) found that for many centuries in England nearly half of all GDP was required to get the energy required to run the other half of the national economy.

Thus this book is about the importance of energy as the world's master resource and also how all life, including humans, have to invest a lot and then get a good return on their investment. Plants, animals and humans themselves who do not invest energy wisely in getting more energy are dead, extinct or exist only as "collapsed" societies.

Therefore investments need to be understood within a broader context, including biological, economic, and energetic. Biologists understand that an organism must meet its own energy requirements for maintenance and generate a surplus before it can invest in reproduction. Likewise economists usually divide what we do with the wealth we have or, especially, produce (i.e., GDP) into two broad categories: consumption and investments. Investments cannot be made into new enterprises until maintenance requirements are met. Hence we, meaning ourselves personally but also a business or a nation, can either spend our wealth or invest it. Basically, it is a matter of time: do we want to enjoy our wealth now or potentially more in the future. As such it is a "zero sum game," meaning that there is an *opportunity cost* to spending your money now: if you spend it now you cannot invest it for the future. For the average person this represents a dilemma, as anyone who has saved to go to college knows.

For monetary investments it seems like a good deal: a wise investment gives you more total money than if you simply spent it now. But from the energy perspective it is a little different, for when you spend money you are also spending energy. The money can be printed indefinitely, but most of the energy cannot, since by definition fossil fuels, which supply about 85 % of the energy we use and are mostly responsible for our great wealth, are not renewable; once used they are gone forever. And energy is essential to give money meaning. Once we thought that gold was a backing for money. Until 1972 one could, in theory, take in your dollars to the U.S. Government and get in exchange gold (the gold standard). But it was not gold that gave meaning to the dollar, it was the goods and services you could trade it for. And those goods and services required energy to produce, about 6 MJ (megajoules; see Table 3.1) per dollar in 2016. If you print or otherwise generate money without energy to back it up the money loses value. Hence we can think of money as lien (a promissory note) on energy. Thus if we are talking about monetary investments we are talking simultaneously about energy investments. Therefore opportunity costs also applies to energy opportunity costs; energy used for one activity or investment cannot be used for another.

Thus the concept of Energy Return On Investment (EROI). For any "progress," indeed simply sustaining any life process requires a continual influx of energy. This concept, often termed "maintenance metabolism", applies to fish in the sea, yourself and your civilization. The reason for this, developed more fully in Chaps. 3 and 6, is that there is a relentless tendency for the molecules that make up an organism (or a city) to degrade to a more random state. Organisms (and governments and private entities) must continually invest energy in keeping molecules in the very specific structures that life requires if it is to continue in its present form. And only after those investments have been met is it possible to have some kind of surplus—whether that is reproduction, evolution or growth or "progress" of a civilization.

Not surprisingly there is a premium on a positive return on investment and especially a high return. If one takes a broad view of evolution any time a new technology with a high energy return happens along in the evolution of life there will be an explosion of life forms using this technology. For example, before some 500 million years ago there was no free oxygen in the atmosphere or the seas. Life was abundant and diverse, but none of it used oxygen (as we do) as a terminal electron acceptor because there was no oxygen available. Life operated on fermentation, generating energy-rich alcohol as an unusable by-product. But once land plants evolved and generated free oxygen as a by-product of photosynthesis, other organisms were able to utilize this oxygen and increase their own utilization of their own food (such as plant sugars) by about a factor of 4. Hence, once the technology of using oxygen was "worked out" through the development of mitochondria, it spread very rapidly throughout the animal kingdom, and also to plants and many microbes. This greater ability to use energy by animals in turn led to much more active and mobile lifestyles, investing existing energy into obtaining much more. Likewise as human societies evolved new energy-capturing technologies (such as agriculture, wind powered sailing ships, and fossil fuels) the technology spread rapidly. Those species and humans who understood or happened upon the new energy-rich opportunities tend to be those (or their descendants) who are with us now.

Humans have always invested their own energy into the requirements for life, originally through hunting and gathering, then through agriculture, then through various industrial processes, and always through conquest and exploitation. Today, this investment process has been institutionalized in Western civilizations through capitalism, including industrial development, colonialization, war and "free trade." This is a very powerful process because, in part, it has so closely combined the expansion of the exploitation of fossil fuels with the economic interests of many of its participants at every level, but especially the top, of the economic ladder. It allows enormous concentrations of wealth and power that in turn reinforce the wealth and power of its participants. There are many in the world today who wish to decrease the power of this process which we examine in Chap. 12.

I find it curious that the generality of this relation—energy as master enabler and EROI as a key determinant—throughout much of the living world has not received more attention and emphasis. But for this it to be understood and appreciated humans need first to understand what energy is and how energy works. This is developed in the next chapter.

Reference

King, C.W., J.P. Maxwell, and A. Donovan. (2015). *Comparing world economic and net energy metrics*, Part 1: Single Technology and Commodity Perspective. Energies 2015:12949–12974.

Chapter 2
The Development of the Laws of Thermodynamics

In order to understand what energy is, it is necessary to understand the laws of thermodynamics. The most important step in this process is unraveling the relation of heat to other forms of energy. Thus we start with an historical perspective of how the human understanding of heat and energy evolved from mystical to increasingly scientific. Much of the interest in heat and energy was initiated by the astonishing success of James Watt's steam engine in the middle of the eighteenth century. While it was obvious that the engine could translate heat from coal into very powerful and useful mechanical work, the underlying reasons as to how were mysterious to many.

2.1 The History of Our Understanding of Energy[1]

In 1750, humans had essentially no clue as to what energy was, how it operated, how it was transformed, and how it was related to work or investments. But in the following century a series of remarkable discoveries and experiments, mostly from Scottish, French, and English scientists, uncovered the essentials of energy and its relation to investments. First were the remarkable discoveries of Isaac Newton—which actually took place 100 years earlier. Newton discovered the three laws of motion (every object remains in its state of motion unless acted upon by an outside force; acceleration is force divided by mass; for every action there is an equal and opposite reaction) and in the more than 350 years since then no fourth law has been discovered! Newton showed that the behavior of matter, which earlier had appeared as chaotic and unpredictable, actually followed a very few mathematically predictable rules. And they were universally applicable, that is they could be repeated again and again by different people in different countries and expanded to other

---

[1]For a much more detailed historical and physical development see: Ayres, Robert. 2016. Thermodynamics, wealth production, complexity. Springer, New York.

© The Author(s) 2017
C.A.S. Hall, *Energy Return on Investment*, Lecture Notes in Energy 36,
DOI 10.1007/978-3-319-47821-0_2

scales or applications. Nature, or at least some of it, did not depend on the whim of deities but regular, understandable, and measureable physical laws. Newton also derived the law of universal gravitation and wrote critically important books on optics, meanwhile (at the same time as Liebnitz) inventing calculus, the mathematics of change. Nevertheless by his own admission he did not understand economics and he lost most of his money on an ill-advised investment scheme. Nor did he understand much about energy beyond the laws of motion.

Most of the additional revolution in understanding energy took place during 1750–1850 (although continuing to some large degree though the next 100 years and even to now). This occurred during a time when there was a tremendous change in the outlook of thoughtful people, often called "the enlightenment," which took place simultaneously and was probably required for the understanding of energy to go forward. The enlightenment movement started in Scotland in roughly 1750 and was soon found in England and France too, then throughout Europe and the American colonies. The essence of enlightenment was a repudiation of the authority of past texts and thinkers (including religious and philosophical sources) as a source of knowledge and an increasing respect for individual thought, observation, measurement, and empiricism. In short the new "religion" was science and the scientific method and all that could be gained by its use.

As of 1750 essentially no one understood energy as a concept, although many people certainly understood many of its practical consequences. They knew plants needed sunlight for growth and that fuel wood could be used to do many useful economic things such as cooking, or making metals and cement. Additionally they knew sufficient practical information to build steam engines and other coal burning machines. But the concept of energy itself was tied up in confusion, often mystical, about its actual form or substance, because energy cannot be seen or felt, but only its effects. A fire heated an iron rod by adding a substance to the rod. Plants and animals as human food were ascribed to special deities rather than as the energy captured from the sun and released from the destruction of chemical bonds generated earlier, and the formation of new lower energy bonds with oxygen. In ancient times fire was thought of as a basic substance (as in Earth, Air, Fire, and Water). As in most other things in their life that they did not understand, most ancient people attributed what we now call energy, or at least some aspects of it, to a god or gods: the sun, of course, was worshiped by many cultures who understood clearly its importance for their food and warmth, but there were many other energy gods: Promethius, Haphaestus, Pele, Vesta, Hestia, Brigid, Agni, and Vulcan to name a few. These people had no possible way to see that there were common concepts linking the fire resulting from burning wood to the sun, nor could they understand that so many other processes that they also attributed to different gods (wind, rain, agriculture, the existence of wild creatures, and so on) were in fact connected to the sun.

The knowledge that a sharp sixth grader today has about energy and science in general would be far beyond what the most learned person would understand four or even two hundred years ago. How could people possibly understand energy if they did not have any concept of photons, chemical bonds, oxygen, or chemical transformations? How could they possibly understand that the growth of plants, the

2.1 The History of Our Understanding of Energy

work of a horse, the erosion by water, the heat generated by fire and their own exertions had a common mechanism tying them all together? How could they understand that heat was a kind of energy, one related to these other things? To them they were independent entities. With the enlightenment came astonishing progress in understanding how the world works. It was driven by a relatively few thinkers and craftsmen who began to leave behind "experts" as a source of knowledge, be they philosophical, religious or even early scientific, and instead trusted more their own thoughtful observations and the development of the scientific method. It is a remarkable story, one that is behind our unprecedented wealth today and one that may eventually, if not fully understood, undermine all that we have accomplished in the name of civilization.

2.2 Heat Energy

The most essential aspect of energy needing explanation was heat, which we now understand as the rate of movement of molecules within a substance. (Technically, the rate of movement determines the temperature, and heat is defined as "the disordered transfer of energy"). The prevailing view of energy in 1800 or so was that heat was a colorless, weightless, substance, generally called phlogiston, that could flow from one entity to another. An iron rod put into a fire would gradually heat up and the heat could be passed to another iron rod (or a finger!) that touched it. The most logical explanation was that phlogiston (or calor, from the Latin word for heat or warmth) would flow from the fire to the first and then the second iron rod. This concept apparently was based on Isaac Newton's (certainly an authoritative figure at the time) belief that heat was an indestructible substance with mass. The first substantial experiment to challenge caloric theory arose from work undertaken by Benjamin Thompson (also called Lord Rumford), a military engineer who invented many practical energy-related devices including the double boiler and a better fireplace. In the 1790s, he watched carefully while recently cast cannons were having a hole bored in them for the projectile. This was done by horses walking in a circle, which turned a sharp drill that drilled the hole into the cast iron cannon. The cannon and drill were immersed in water to keep the cannon from overheating, and the water would boil after several hours drilling. People thought that phlogiston moved from the cannon or the drill into the water. But Rumford observed that if the drill was blunted that the water could be made to boil indefinitely—in other words the cannon did not run out of phlogiston. He surmised that the heat did not come from phlogiston but came instead from the effects of friction causing the materials in the cannon and then the water to move more rapidly. Thus mechanical energy in the drill was being converted into molecules moving more rapidly, i.e., heat.

During the same period there was a lot of practical work carried out to turn coal into steam for running steam engines. Early engine makers were impressed with the work potential, but also the very low efficiency by which it was generated. Newcomen's early engines operated at only about 1 % efficiency (partly because the cylinder had to be cooled to condense the steam during the power stroke, and then reheated when the steam was reintroduced, so the energy in the steam was used mostly to reheat the cylinder rather than move the piston). Even James Watt's great improvements resulted in engines of only about 3 % efficiency. In other words, most of the heat obtained from the coal did not generate mechanical motion of the engine, but just useless heat, and thus the engines were expensive to run, needing a lot of coal. The mechanical work returned on the coal invested was very poor. So practical engineers were wondering why engines were so inefficient, and phlogiston theory had no particular answers.

Also during this time Joseph Black, Antoine Lavoisier, and others were generating important new ideas about latent heat energy (e.g., the energy to change water at 100 °C to steam at 100 °C.) and various considerations of heat related to chemical reactions, as well as the existence and nature of oxygen and carbon dioxide. Many people were beginning to understand nature much better, quantitatively and thoughtfully, but there was no particular theory to tie all these observations together.

2.3 Developing the Laws of Thermodynamics

Our most important concept in understanding how energy investments, and energy more generally, operate was the development of the laws of thermodynamics, which occurred mostly in France and England during the second quarter of the nineteenth century. The first major step integrating knowledge about heat with a scientific perspective was undertaken by a young French military engineer named Sadi Carnot (Fig. 2.1). His 1824 book *Réflexions sur la puissance motrice du feu et*

Fig. 2.1 Sadi Carnot (1796–1832): the "father" of thermodynamics

2.3 Developing the Laws of Thermodynamics

sur les machines propres à développer cette puissance (*Reflections on the Motive Power of Fire*), is generally considered the starting point for the modern science of thermodynamics ("therm" means heat and "dynamics" motion or change). Carnot sought to answer two questions about the operation of heat engines (such as steam engines, but he wished to generalize): "Is the work available from a heat source potentially unbounded?" and "Can heat engines in principle be improved by replacing the steam with some other working fluid or gas?" The most important part of the book was a theoretical development for an idealized engine that could be used to understand and clarify the fundamental principles for any heat engine. Carnot defined "motive power" as the expression of the *useful effect* that an engine is capable of producing defined as *weight lifted through a height*. Today we call this work. In his time, and from the perspective of the times in which he lived, this idea of a weight lifted through a distance made a great deal of sense because most of the work needing to be done, at least initially, was to lift a quantity of water out of a mine. Carnot made the analogy that work can be done as heat "falls" from a warmer to a cooler temperature analogously to water falling from a higher elevation to a lower elevation, potentially operating a water wheel in the process.

Perhaps the most important contribution Carnot made to thermodynamics was his abstraction of the essential features of the steam engine into a more general and idealized heat engine. This resulted in an idealized thermodynamic system on which exact calculations could be made, which avoided the complications of the crude features of the contemporary steam engine. By idealizing the engine, he could arrive at clear and indisputable answers to his original two questions. He showed that the efficiency of this theoretical engine depends on only two things: the temperature of the source of heat and the temperature of the sink into which it is placed after use. This was later formalized in the equation:

$$E_{max} = (T_1 - T_2)/T_1,$$

where: E_{max} is the maximum efficiency of the work possible (such as lifting water out of a well) by an ideal (i.e., frictionless) heat engine, T_1 is the absolute temperature of the source of energy (for example, steam exiting a boiler) and T_2 the sink, i.e., a river or the atmosphere. The Carnot efficiency by definition cannot exceed 1, and in actual operation with modern machines is usually on the order of 25 (diesel and gasoline engines) to 50 % (gas combined cycle turbines). In general, it is higher with a greater difference between T_1 and T_2. In Carnot's rather amazing words: "The production of motive power is therefore due in steam engines not to actual consumption of caloric but to its transportation from a warm body to a cold body." He devised a graphic representation of an ideal heat engine that, using assumptions of, e.g., no heat loss to the machinery, gave a maximum efficiency that no engine could (or has) exceed (Fig. 2.2).

This equation explains why, despite the vast amount of heat stored in, for example, the surface of the North Sea in summer, so little work can be done from it: the difference between the surface temperature (30° C) and the deepest water (2° C) is too small compared to, say the temperature difference in an oil fired power plant,

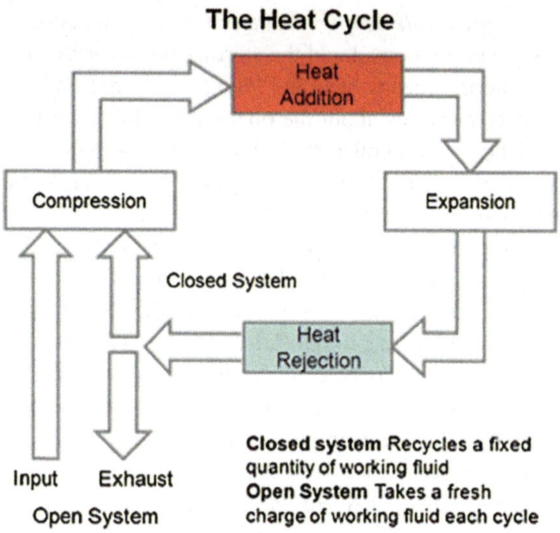

The Heat Cycle

[Diagram: A cyclic flow diagram showing Compression → Heat Addition → Expansion → Heat Rejection → back to Compression. Labeled "Closed System" in the center. Arrows at the bottom indicate "Input" and "Exhaust" for "Open System".]

Closed system Recycles a fixed quantity of working fluid
Open System Takes a fresh charge of working fluid each cycle

Fig. 2.2 Diagram of a basic heat engine, such as a steam engine. Work is done during the expansion phase, such as when hot steam or exploding gasoline pushes a cylinder and converts high-grade heat energy into mechanical work. Some of that heat is recoverable, but there is always some part that cannot be recovered. Both must be rejected in the exhaust system (from http://www.mpoweruk.com/heat_engines.htm)

where temperatures at the turbine entrance may reach 817 °C and the cooling water which might be from 6° C (winter) to 17° C (summer). Entropy is the name usually given to low grade, low-temperature heat unavailable to do further work. We will give it other definitions later.

Similarly the change in entropy of a system is determined by its initial and final states. In theory the (ultimate) final state is absolute zero, as temperatures cannot go below that. Thus the quantitative estimate of S, the entropy (or perhaps more accurately the amount of energy that cannot possibly be used again) after an action, is

$$S = (E_{init} - E_{fin}) / T_{abs}$$

where S (or technically dS) is the generation of entropy from a process, and $E_{init} - E_{fin}$, is the difference in temperature of the heat from the start and ending of the process.

The most fundamental equation is that the total amount of energy available to do additional work after a given transformation can be calculated as

$$Wpot = E_{init} - E_{fin} - Sv$$

where Sv is the entropy generated during the process. Carnot's equation has many important implications today. To generate a more efficient jet engine, ceramics engineers build turbine blades that can withstand hotter and hotter gases upon them,

2.3 Developing the Laws of Thermodynamics

and power plants operate slightly more efficiently in winter, when E_{fin}, the river water used for cooling, is colder. But Carnot as of 1824 still seemed confused about the nature of heat, as his publication is broadly consistent with the phlogiston theory. Tragically, and in the fashion of the times, his personal effects were burned after he died, including nearly all of his notes and developing papers. The few notes that survived indicate that he was beginning to understand heat as molecular motion and they anticipated most of the groundwork for the first law of thermodynamics. Unfortunately, they remained undiscovered and unpublished until 1878.

The next critical step in the understanding of thermodynamics was made by James Joule, a brewer with a very curious mind. He measured again and again what we now know as the mechanical equivalent of heat. In 1845, Joule reported his best-known experiment, which measured the mechanical equivalent of heat by taking a pulley and rope, attaching a weight to one end of the rope and wrapping the other end around a shaft that went into an insulated water chamber where it operated a paddle wheel (Fig. 2.3). As the weight dropped (doing so many kilogram-meters of

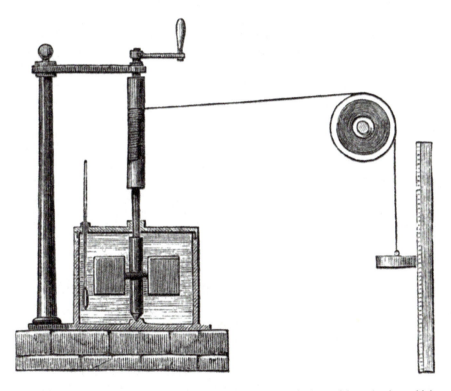

Fig. 2.3 Joule's basic machine to derive the mechanical equivalent of heat (or it could be expressed as the heat equivalent of mechanical work). The known weight would drop a specific distance meanwhile turning a paddle wheel immersed in water, and that mechanical energy would be translated into an increase in the temperature of the water

mechanical work—think again of water being raised from a mine) the temperature increase within the chamber could be measured with a thermometer. He showed that the potential gravitational energy of the elevated weight was equal, after the weight was dropped, to the total heat added to the water by friction with the paddles. By doing so Joule found that one newton-meter of work (or 7.2 foot pounds) was equivalent to 1 Joule of heat energy. This led to the understanding that heat was a form of energy and, eventually, the theory of conservation of energy.

Neither Joule's publication nor Carnot's *Réflexions* had much impact initially: Joule's careful work was not given its due initially because he was a businessman, not a "gentleman," and Carnot's work, even after being modernized by Emile Clapeyron in 1834, still received relatively little attention for another decade. But then in 1845 R. Holtzman in Germany and William Thomson (Lord Kelvin) in Scotland, both working directly from the *Réflexion*, started publishing a series of papers that both extended and confirmed Carnot's results. Subsequently Rudolf Clausius and Lord Kelvin together derived from it explicitly the concept of entropy and the second law of thermodynamics.

In 1850, Clausius gave the first clear joint statement of the first and second laws of thermodynamics, abandoning the caloric theory, but preserving Carnot's principle. We then understood (or at least those who read these works did) how heat and energy and work were related. With this the operation and meaning of much of the physical world became obvious, and opened up enormous applications, especially through the industrial revolution—which has generated so much human welfare as well as misery.

Thus the laws of thermodynamics, which today seem so obvious, were in fact developed very, very slowly. Key issues were missed: in 1600, the English philosopher and scientist Francis Bacon surmised: "Heat itself, its essence and quiddity is motion and nothing else." But this was not generally accepted, even by the scientists studying it, until after 1850. Savory, Watt, and others working with boilers and steam engines were thinking all the time about converting heat to work, and certainly were aware of friction, but did not understand motive energy and heat as a two way street. Carnot's work, too, took decades to take hold. But once these ideas were summarized as the laws of thermodynamics they became relatively clear and widely accepted.

2.4 Understanding the Laws of Thermodynamics Today

Thermo means heat (or energy) and *dynamic* means change. *Thermodynamics* is the study of the relations between motion and heat and the transformations that take place as energy or fuels are used to do work. *Work* occurs when something is moved, including, for example, a rock or your leg lifted, a car driven, water evaporated or lifted up in the atmosphere, chemicals concentrated or carbon dioxide transformed from the atmosphere into a green plant. There are two principle laws of thermodynamics, called the *first law of thermodynamics* and the *second law of thermodynamics*. Quite simply the first law says that energy (or for some special considerations energy-matter) cannot be created or destroyed, but only changed in form. Thus the potential energy found in a gallon of gasoline in a car's fuel tank is transformed through driving into other forms, primarily heat associated with overcoming friction to create the momentum of the car, as heat dissipated by the radiator or where the tires meet the road, or in the increased potential energy of the car if it is driven to the top of a hill. Most of that original energy is dissipated into the environment as low-grade heat—from which it is essentially impossible to get any additional work. Technically you could capture that waste heat and then use some of it, but it would require the use of even more energy to do so. Some fraction of the work done can be used again, for example the automobile could be rolled back to its original downhill position using the force of gravity. But less energy would be recovered than was originally in the gallon of gasoline, because some of it has been dissipated into low-temperature heat.

The second law of thermodynamics says that all real-life processes produce low-grade heat, often (but not always usefully) called entropy. At every energy transformation some of the initial high-grade energy (that is energy that has potential to do work) will be changed into low-grade heat barely above the temperature of the surrounding environment. In other words, the first law says that the *quantity* of energy always remains constant, but the second law says that the *quality* is degraded over time. The practical meaning of this for people and economies is that, with the exception of the energy from the sun, it is always necessary to find additional energy resources to construct and maintain whatever structures we have, including plants, animals, houses, cars, civilizations, and ourselves. The implications of this have had overwhelming impacts upon all human enterprises and histories, and constitute the remainder of this book.

To the best of our knowledge every action that occurs on earth, or elsewhere, is subject to the laws of thermodynamics. The only possible exception is that the law of conservation of energy needed to be expanded to the law of conservation of mass-energy to encompass nuclear reactions (in a star, nuclear bomb or nuclear power plant). This is because mass can be converted to energy (and the converse) according to Einstein's famous equation: $E = MC^2$, which says that under special circumstances energy created equals mass times the speed of light squared (a very large number). In other words in a nuclear conversion a very small amount of mass can be converted into a huge amount of energy, although this can take place only under very special conditions.

2.5 Implications of the Second Law

One important implication of the second law is that heat cannot pass spontaneously from a colder to a hotter body—energy always must dissipate from a warmer body to a colder body, and go to a less organized and less useful state in the form of waste heat that cannot be used to do useful work. A second, somewhat confusing concept, related to the second law is the "Law of Entropy." Entropy is a measure of disorder or randomness in a system. With respect to the second law it might better be called a measure of the amount of energy that, during an energy transformation, becomes no longer available to do work. It is associated with the second law because energy must be expended to create and maintain order against disorder. Most fundamentally order is the non-random structure of molecules, such as a sandwich, an automobile, or a person. Thus to have order (i.e., life, or societal infrastructure) energy investments must be undertaken constantly, and in the process some of that energy must be degraded into heat and hence lost for other uses. One way to think about it is that organisms, including yourself, are centers of order, as determined by the DNA inherited from your biological parents. To maintain this order requires energy captured from the environment (as food) and invested in the multitude of processes that maintain the organisms in the very non-random pattern that represents an organism. In the same way a refrigerator takes in energy from outside itself (electricity from the wall socket) and uses it to maintain order within (i.e., the tuna sandwich). In the process some of that invested energy must be degraded into heat (at the back of the refrigerator), so that disorder is created to maintain centers of order. The universe as a whole is said to be more disordered as a result, but order is maintained where it is desired—that is where natural or cultural selection finds it useful. One result, of considerable concern to a few (but not me), is that since, apparently, the "big bang" at the origin of the universe high-grade energy, such as very high temperature matter in stars, light, and so on is being slowly degraded to low-grade heat so that the background temperature of the Universe has increased to barely above absolute zero.

In my opinion there has been too much emphasis placed on entropy, or disorder. What is important to life and civilizations is not disorder, which is ubiquitous, but order. Order should be our focus because it represents the maintenance of biological and economic structure, of very specific molecules in an organism, well-placed bricks in a city or tuna fish in a sandwich. The price of maintaining this order (sometimes called negentropy or in a different context, exergy) is extracting and using energy from the environment, a common example being the use of fossil fuels, usually associated with financial investments from private sources or the proper use of taxes. A by-product of the use of this energy is the degradation of some or most of it to low-grade heat. That low-grade thermal energy is lost to us forever, but fortunately there is plenty more high-grade stuff for us to use so we seldom give that loss any further thought. So entropy is important, yes, but what is of greater importance is the order, and the investment of energy into maintaining the order needed for life and civilizations, and for how long we will be able to do that. The amount of heat or disorder added to the universe, or even the Earth, from the process of life-maintaining order is trivial and has little or no importance.

The implications of the Second Law are profound. Systems that involve energy transformations are unidirectional and irreversible (unless additional energy is used). They lose energy as low-grade heat and may increase the order of some parts of that system. This is sometimes called time's arrow: formally "in an isolated system, one never observes a spontaneous transformation of disorder into order, one always observes a degradation of order into disorder." Another way of stating this is that if left to themselves, things get more disordered over time (think of a young

2.5 Implications of the Second Law

child playing with blocks—or for that matter your closet). In fact if you are frustrated by the fact that the world seems constantly disordered and working against you—Where are my keys? Where are my glasses? Why has the washing machine broken?—it is just because it is true, the natural tendency of all things is toward disorder, and if order is to be maintained energy must be invested in that. A corollary to this is that the "entropy" of the universe increases over time, as apparently has been the case since the original "big bang" which created much order of a sort, although that is beyond the scope of this book or its author! Since

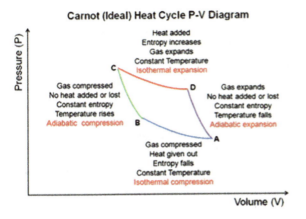

Change of State	Carnot Heat Cycle Processes

A to B   Reversible isothermal compression of the cold gas. Isothermal heat rejection. Gas starts at its "cold" temperature. Heat flows out of the gas to the low temperature environment.

B to C   Reversible adiabatic compression of the gas. Compression causes the gas temperature to rise to its "hot" temperature. No heat gained or lost from system.

C to D   Reversible isothermal expansion of the hot gas. Isothermal heat addition. Absorption of heat from the high temperature source. Expanding gas available to do work on the surroundings (e.g. moving a piston).

D to A   Reversible adiabatic expansion of the gas. The gas continues to expand, doing external work. The gas expansion causes it to cool to its "cold" temperature. No heat is gained or lost.

If the heat cycle is operated clockwise as shown in the above diagram, the engine uses heat to do net work. If the cycle is operated in reverse (anti-clockwise), it uses work to transfer thermal energy from a cooler system to a warmer one thereby acting as a refrigerator or a heat pump.
(Modified from http://www.mpoweruk.com/heat_engines.htm)

Fig. 2.4 Carnot's basic diagram showing the pressure (P) and volume (V) relations of an ideal heat engine. Work is done from location 1 to location 2, as volume expands and pressure decreases (such as when a piston is being pushed on the power stroke)

our sun will be operating more or less as it is now for additional billions of years it is certainly not worth losing any sleep over.

Organisms, and sometimes entire ecosystems, are called "self-organizing systems" because they maintain their own organization essentially by themselves, as a result of natural selection for genetic codes that cause them to extract energy and use it to generate and maintain matter into the incredible machines that are living organisms and ecosystems. Additionally life forms are called "autocatalytic" because they catalyze their own growth: exploitation of energy from the environment allows them to build structure that allows even more exploitation and growth, until the limits of available energy are reached. One might say that life breaks the second law, as organization is generated seemingly spontaneously. Yes, this happens, but only through the degradation of much of the free energy of the Earth-Sun system to heat, so the second law remains intact. There is an energy tax for all activities, imposed by the second law and various inevitable and even desirable inefficiencies, and we cannot escape it (Fig. 2.4).

2.6 The Second Law and Efficiencies

It is critically important to understand that every energy transformation comes at an energy cost and cannot be 100 % efficient. Ecologists understand the importance of the Second Law because it applies to every living organism and ecosystem alike. Also worth noting is how stunningly inefficient many biological processes are from an energy perspective. Photosynthesis converts at most about 2–4 % of sunlight into plant biomass. The rest is lost as heat. Similarly, herbivores typically convert only 10–20 % of the energy in the plant biomass they eat into energy bonds in their own flesh. The transfer efficiencies, which ecologists call "ecological efficiencies," are roughly similar for each succeeding "trophic" or food level. By the time a top carnivore such as a tuna fish or smallmouth bass eats its prey, only a very small fraction of the original energy from the sun remains in the food chain (Fig. 2.5). The rest has been dissipated as heat, mostly from maintenance metabolism along the way. This sort of analysis is helpful to us in understanding how ecosystems and human systems alike function. Similarly, the understanding of transfer efficiencies needs to be considered if we want to make sensible energy policy.

Why is this concept relevant to energy policy? The answer is simple: the more transformations that occur in producing or using a given energy source, the more of the energy will be lost as waste heat. Some transformations are more efficient than others, so we might consider maximizing our use of those. For example, internal combustion engines are notoriously inefficient at converting their fuel's chemical energy to mechanical energy, usually at from 20 to 30 %. Engine friction and waste heat are major factors accounting for low efficiency, both of which relate to the Second Law. On the other hand, electrical engines are much more efficient, approaching 90 % efficiency in converting electricity (a higher quality fuel) to mechanical energy. However most of our electricity comes from burning fossil

2.6 The Second Law and Efficiencies

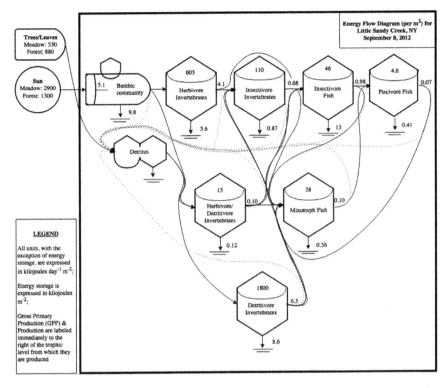

Fig. 2.5 Diagram of the food chain of a small stream in upstate New York for a day in September. 2012. Energy stocks and flows are estimates of mean energy flow using symbols from H.T. Odum. *Bullet-shaped* modules represent autotrophs, hexagons represent heterotrophs. *Dashed lines* represent indirect energy flow to detritivorous invertebrates. *Numbers in hexagons* are biomass values and *numbers on downward pointing arrows* are respiration values. GPP and production values are on the *horizontal lines/arrows* to the right of the trophic level from which they are produced. Of the 2000 MJ of sunlight and 600 MJ of leaves that enters a mean square meter of the stream in a day only about 1 MJ makes it to the top carnivore, in this case piscivorous fish such as a small mouth bass (From Hall et al. in revision)

fuels at roughly 30–40 % efficiency. Thus if we started with oil and used it to drive a train the efficiency would be, say, 0.32 or so. If we used that oil to make electricity which then drove the train's electric motors the overall efficiency would be 0.4 times 0.9, or a total efficiency of roughly 0.36, a little better than burning the oil directly, but not much. Consequently most modern locomotives in the U.S. are "diesel electric," burning diesel to make electricity and then using the electricity to drive the train.

Now that we (and the scientific world of the early nineteenth century) have a practical definition and understanding of thermodynamics we turn to a more general consideration of energy which is much more informative with the relation of energy and heat defined.

Reference

Hall, C. A. S., F. Knickmeyer, A. Wiegman, A. Brainard, A. R. Diaz, C. Huynh and J. Mead. (in revision) A class exercise for systems ecology: synthesis of stream energetics and testing Allen's paradox. Ecological Modeling.

Chapter 3
About Energy

Most people do not think much about it, but energy and its effects are pervasive and ubiquitous in our life. Most of us eat three meals a day to get the energy to run ourselves, but we do not usually think about this in terms of energy. We fill our car with gasoline but rarely think about its energy content, where it came from or whether our children or grandchildren will be able to fill their cars with gasoline. We watch athletes compete in amazing ways that are about the quantity and skill with which they are able to use their own personal energy but again do not think too much about the way food energy is transformed into the amazing feats or the underlying physics of the process. For example, from 70 to 80 % of the top three place winners of the most competitive long distance running contests, such as the Boston Marathon or the summer Olympics, are from the country of Kenya, even from a single tribe, the Kalenjin, of whom there are only 4.4 million. But they have a special build with relatively thin (but powerful) lower legs. The best sprinters, on the other hand, are often directly or indirectly from West Africa and tend to be characterized by heavy, powerful calves. It seems that physics, and Newton's second law, favors not having much mass to push forward for long distance running, and extremely powerful lower legs for a quick start. One can conjecture that long distance running was favorable for running down antelope (or escaping lions) in dry, open East Africa (see e.g., Glaub and Hall in press) and getting a quick start was useful in wet, heavily forested East Africa. The point is that everything we do, and are capable of doing, depends on energy and the specific genetic or technical means of using it.

Nevertheless even when we speak of energy commonly in everyday life we are usually not talking about energy itself but rather the fuels or energy carriers, entities such as gasoline, electricity or food that contain or *carry* energy. Why is this so? If energy is so important then why is it not more generally understood, taught, and appreciated? The answers are complex. One important reason is that there tends to be a proclivity to give people far more credit for controlling events in their lives than a careful examination of the facts warrants. In other words humans tend to give explanations based on human actions or will when in fact often there are better energy-based explanations. For example, the credit for winning the second World War is normally given to Allied military and political leaders rather than to the (arguably greater) importance of the available fuel and industrial capacity of the United States, a fact well understood by, for example, the great Japanese Admiral Isoroku Yamamoto who had studied in the United States, visited its shipyards and advised against attacking Pearl Harbor (but then, over ruled, carried out the Pearl Harbor attack brilliantly). Today, for another example, most people give the credit for economic growth to some ideology, political leaders, or economists, when often the main reason is related to the availability (or not) of cheap energy. In contrast Hamilton (2009) found that, for example, most recessions in the last 50 years in the United States followed a sharp increase in the price of oil. We will dig into this issue in much more detail later.

A second reason people do not think much about energy in their daily lives is that the energy used to support ourselves, our families, and our economic activity generally is combusted at some other location and by other people, or by quiet, automatic machines whose fuel tends to be relatively cheap and invisible. After all, coal, oil and gas, our principal sources of energy, are basically messy, smelly, dangerous, and unpleasant materials. The electrons in the wires we plug into the wall, derived far away from a dam, or the combustion of these fuels, are invisible and noiseless as they go about their prescribed and automatically regulated work. The food energy that we need to fuel ourselves surrounds most of us abundantly, is relatively cheap for most, and the chemical and embodied energy within it is disguised by our attention given to flavors and preparation or viewed negatively as calories as our collective waistlines expand. Houses are kept at comfortable temperatures by nearly silent machines hidden in our basements, using fuels delivered seamlessly in wires or pipes. Society has gone to great lengths to isolate most of us physically and intellectually from the energy sources upon which our food, our comfort, our transportation, and our economy depend. It is convenient to ignore energy in part because many facts about it are uncomfortable to know or experience. To some degree this is changing as the world appears to be warming and the combustion of fossil fuels is largely held to blame. But mostly our relation to energy is "out of sight, out of mind."

Well-established laws of physics and chemistry define and characterize the specifics of the role of energy in the natural world. Energy and its laws not only control all nonliving physical–chemical phenomena, they equally define and constrain life, which can exist only within and according to these laws. Energy controls and sets limits to the structure and dynamics of all levels of biotic systems, including the maintenance of a habitable biosphere, the origin and sustenance of biodiversity, natural ecosystems and the evolution of human civilizations and their economies.

3.1 What Is Energy?

But what is energy? Defining energy turns out to be more difficult than what one might think. The high school physics definition of energy "The ability to do work" does not take us very far, for what is ability? And what is work? What is the stuff (or non stuff) that allows this to occur? Robert Romer wrote a good physics textbook which was about using energy concepts to understand all the conventional material of physics because "all physics is about energy." Yet even he admitted that he was unable to give a satisfactory definition of energy. He said we can see energy's effects, we can measure them, but we do not really know what it is. Physicist Jacques Treiner recognizes the difficulty in defining energy precisely. He says "Energy, in the scientific context we are dealing with, is an abstract notion, of mathematical nature, which allows us to quantify the transformations of matter." Usually we detect energy containing or transforming materials (food or petroleum and oxygen) and energy being used because something is moved: a car, a basketball player, chemicals against a gradient, and so on. Hence energy can be thought of, not quite precisely, as that which causes motion. For our day-to-day experiences, energy is mostly found associated with either photons coming from the sun or the oxidation of fuels such as wood or food or oil that generates work (i.e., moves something) at some point in space and time. These are things we can experience and understand pretty well, even if the physicists cannot define them exactly, at least in a way that most of us can readily grasp. One of the things that makes energy easy for me to understand fairly well is that, having measured biological energy a great deal, I am impressed with the general sense and repeatability of the measurements. But I understand that I am not measuring energy itself, but rather its effects.

On a broader scale energy, or more precisely the transformations energy makes in matter, runs the world, figuratively and literally. Its effects are pervasive, relentless, and all-encompassing. While one cannot see energy, we can readily see its effects in transforming matter. What does this mean? It means we can readily and frequently observe the effects of energy transforming matter: you will your feet to go up the stairs and they do, lifting your body mass and generating heat. Your puppy shows incredible energy on a walk. Plants grow in sunshine and pretty much disappear when burned, meanwhile emitting light and heat. We can drop a stone from the top of a tower and watch it fall. We can step on the gas and a massive chunk of iron will be accelerated rapidly. Thus we mostly do not observe energy directly but rather examine the effects of energy in transforming matter. Most of the energy around us is bound up in some kind of matter, contained in elevated rocks or water, the chemical bonds of wood or coal or food, warmer spaces inside our homes or other buildings, and so on. And we can readily observe bound energy being transformed into motion by watching cars zoom down the highway.

Physicist Jacques Treiner has thought a lot about energy. He says we do not live on energy. *We live on transforming matter around us (and, C.H. adds, extracting the energy within it).* One way of visualizing the notion is to ask *how much matter has to be processed in order to deliver or consume a given amount of energy*? The answer depends of course on the type of energy and the interactions involved in the process. The more intense the energy interaction involved, the less matter will be necessary.

For example, an energy transaction of 1 kWh (equivalent to 3.6 MJ) corresponds to (approximately):

- 10 tons of water running down a medium-sized hydroelectric plant.
- The combustion of 1 kg of fossil fuel.
- The fission of 1 mg of uranium in a nuclear plant.
- The fusion of 5 μg of hydrogen in the center of Sun.

Obviously nuclear changes are much more intense than what we are used to.

Coming back to the simple definition of energy as "the ability to do work" still seems somewhat vague and inadequate, for what is work? Does this include heat (well, the molecules are moving faster)? Most simply, but imperfectly to the physicist, work occurs when "something is moved." Thus energy, whatever that is, applied to transforming matter, is what causes motion to occur. Energy is experienced by humans most often as photons (with its associated energy) flowing from the sun or byproducts of the energy associated with those photons. We can see the atmosphere respond to this energy input as summer cumulous clouds rise into the atmosphere. We can watch water slowly disappear from a bowl. We can watch plants in our garden grow day to day as some small part of this photon flux is captured by plants through photosynthesis and stored as reduced (i.e., hydrogen—and energy—rich) chemical bonds associated with carbon. Then we can watch animals grow as they eat plants or other animals built with the energy of the plant, as that energy is passed as electrons associated with hydrogen through food chains to an electron acceptor such as oxygen. Thus we are able to use the energy in a hamburger by oxidizing the reduced matter of the plant and animal tissue therein which initially was obtained by grass (such as wheat) that captured and stored that energy from the photons, and then passed it as chemical bonds to the cow and then to us. Likewise when we drive an automobile we are oxidizing oil that is constructed of high-energy chemical bonds originally made with energy captured from the sun by algae but then processed over some 100 million years of geological pressure cooking.

Although the energy associated with each of these processes may be measured in very different units, they are all interchangeable, and "boil down" to the ability to heat water. So we can generalize that, according to the laws of thermodynamics, all forms of energy share the ability to be turned into heat. This gives all different types of energy a common yardstick by which they can be measured.

3.2 Biology

The biological processes are analogous to that of energy-rich electrons passing through wires in an electric circuit to an electron acceptor (called a ground, the second wire needed in a circuit) and doing work, such as running a motor, in the process. The main difference is that the biotic process is slower, highly controlled by many biochemical microcircuits and usually associated with hydrogen. So, intuitively, we can understand the effects of the energy flows around us in a straightforward way. Electrons get a "kick" (from sunlight, a battery, a generator) which energizes them, and this is dissipated as it returns to its original, or ground state, where it is frequently given another "kick" by a photon, a battery, or generator.

The effects of energy are perhaps most obvious as biological processes related to sports or as fuels to run our vehicles and to heat our homes and other buildings. In general fuels are reduced compounds, where "reduced" means a material rich in hydrogen and energy, but poor in oxygen. They are generally a hydrocarbon (CH_n) where n means some number such as 2, 4 or many) like oil, or a carbohydrate (CH_nO), such as glucose sugar ($C_6H_{12}O_6$), for example food or the alcohol added to gasoline. The "ate" on the end of a word refers to the presence of some oxygen, so that a carbohydrate will have roughly a third less energy than a hydrocarbon per gram, but still enough to be used as a fuel. When a reduced fuel is oxidized, i.e., combined with oxygen, energy is released, the electrons lose their "kick" and the hydrogen is released as water (H_2O) and the carbon as carbon dioxide (CO_2), with both the C and the H atom eventually combined with oxygen. There is no more chemical energy available from these fully oxidized materials.

The basic equation of life is:

$$6\ CO_2 + 6\ H_2O \leftrightarrow C_6H_{12}O_6 + 6\ O_2.$$

The process of respiration is chemically the same as photosynthesis but runs in reverse (from right to left in the above equation). Energy is required to run the equation left to right, and released when it goes right to left.

3.3 Quantity of Energy

One reason people do not appreciate the role of energy in their day-to-day life is the different units routinely used. We are used to measuring food in calories (confusingly meaning kilocalories when capitalized), electricity in KWh, natural gas in therms, engines in horsepower, and liquid fuels in gallons (or liters). How can you compare them? Actually they all measure only one thing, the ability to heat water, and all are intra-convertible simply by multiplying by a constant. 1 kcal (4187 J) can heat 1 liter of water 1° centigrade. The preferred units now are Joules and

Joules per time, whose value and nomenclature are maintained by the International System of Units (SI), along with conversions from other units to the preferred Joules. It is strange that the United States is one of only three countries in the world that has not converted to the metric system, although it is said that gradually the U.S., following Canada, is "inching towards the metric system." The metric system makes calculations enormously easier. If we used Joules for all energy measurements many of the mysteries of energy would disappear. In France, food has its energy content given on the label in kilocalories and kilojoules (4.2 times larger), a step in the right direction.

3.4 Exergy and Emergy

An important consideration, encompassing both quality and quantity of energy, is the amount of work one can get out of a unit of energy. Since some heat is of necessity lost during every energy transformation, it is technically more correct to consider energy not in terms of the amount of heat it will generate but the amount of work it can do, after correcting for the inevitable loss of heat that will of necessity be lost during the use of that energy. Thus many physicists prefer to use the term *exergy* for the amount of useful (e.g., mechanical, electrical, chemical) work that a unit of energy can deliver. The term *anergy* is used for the nonuseful portion of a fuel, that portion that will be changed into low grade heat that cannot be used again (sometimes called entropy). While the exergy in a fuel depends on its particular application and environmental temperatures, it is often in the vicinity of 90 % of the heat value for many common applications with fossil fuels. See Rangel-Hernandez et al. (2016) for a recent application.

The terms *exergy* (considered above) and *emergy* have been introduced by physicists and ecologists to represent some aspects of the quality of energy. Exergy specifies the ability of various energies to do work. Emergy is a less precise but more comprehensive term that includes all of the various energy inputs, including those from the natural environment such as water, to make something, weighed by how many steps each is from the original solar source. We do not consider it much in this book but many believe it gives a powerful way to include the environment much more completely in energy analyses.

3.5 Quality of Energy

Even though there is more coal than oil in the world, there are many aspects of the latter that make it a more desirable energy source. For one thing most coal remaining is in thin or deep seams that are very hard to exploit. Second, oil can do most things coal does (heat water, run trains) but far more things as well-like fueling trucks, tractors, and airplanes. Third, oil is more energy dense. A kilogram

3.5 Quality of Energy

of oil has about 50 % more energy than a kilogram of coal. Fourth, oil can be transported and combusted much more easily and cheaply than coal.

The most obvious example of energy quality is in the food we eat, as what is important to people in addition to energy is type of energy as well as many critical nutrients. The energy contained in corn has obvious utility to us as food, where the energy contained in wood or coal does not. There are many other aspects of food quality. Corn is a versatile and very productive crop. It (or other grasses such as wheat or rice) often represents a substantial proportion of the diet in poorer rural communities because it generates the most food production per hectare. But conventional corn is very low in a critical factor absolutely required for humans: the amino acid lysine. If the corn is fed to a cow then the energy bonds in the corn are transferred to energy bonds in the flesh of the cow, although with very low efficiency as we have described. Cows can make lysine from their food, so protein from a cow has a full complement of amino acids and hence is a higher quality food for humans, at least from that perspective. Many relatively poor people in Latin America (and elsewhere) eat mostly rice and beans. The rice and beans are cheap and they complement each other: the amino acid lysine is missing in rice but found abundantly in beans, while rice is basically carbohydrates, a good energy source, and beans are protein rich. Thus rice and beans provide an excellent diet for humans, although it is still missing one critical ingredient: vitamin C. Fortunately, people who have a rice and bean diet often use chile peppers as a condiment, and such peppers have high levels of vitamin C. Thus cultural selection in humans often appears to be associated with real dietary needs, all of which ensures that generally the energy that fuels humans has the required quality.

Just as people may feed rice or another grain to a cow to get a smaller quantity of higher quality food energy, coal, or oil can be burned to generate a smaller quantity (as measured by heating ability) of electricity. The exceptional utility of electricity gives it a higher quality than the heat available from direct combustion of the fossil fuel. It can be used to do things such as light a light bulb or run a computer that one cannot do with oil or coal. We are willing to take roughly 3 heat units of coal or oil and turn it into one heat unit of electricity because it is more useful to us, and hence is more economic, in that form (the other two thirds of the energy is rejected as heat to the atmosphere or an adjacent water body). We say the *quality* of the electricity is higher than the energy directly available from coal or oil.

There is another important aspect of energy quality, which is the energy cost of getting it, and this will be covered in Chaps. 10 and 11.

3.6 Types of Energy

Energy also comes in many forms or types. First, there are two major categories or what I will call *types* depending upon whether or not that energy is actually being used (i.e., there is motion associated with it and some of it is being turned into heat) versus energy associated with position, i.e., being stored in an elevated mass. *"Kinetic"* is the term we use for energy actually being used at a given moment:

The kinetic energy of a body equals the mass times the square of its velocity:

$$E = 1/2\,MV^2$$

This simple equation means that as you drive faster there is more and more damage that can occur if you crash into a solid object.

Potential energy is stored energy, which might be water at an elevation, electrons in an unconnected battery, gasoline in a tank, and so on. Energy is usually associated with a "gradient" between higher energy and lower energy states, as seen most clearly with elevated water. Potential energy can be transformed into kinetic energy and the converse, although there is always some loss of exergy, the ability of energy to do work, required for any transformation, which is an implication of the second law of thermodynamics. A third type of energy is *embodied* energy, which is the energy once used to make something, such as a chair or automobile. Embodied energy is no longer able to do work. A *primary energy source* (e.g., solar radiation, fossil fuels, or waterfalls) is an energy source that exists in nature and can be used to generate *energy carriers*. An energy carrier is a vector derived from a primary energy source (e.g., electricity, gasoline, or steam) (Murphy and Hall 2011).

Energy also exists in many different *forms*: as gravitational, geothermal, heat, mechanical, sound, chemical, light, and nuclear. The energy in one form can be converted into another but with an inevitable reduction in exergy or potential to do work. The most important energy source for most things going on at the surface of the Earth is the solar energy from the sun which fuels many complex natural processes including life itself. For example, solar energy evaporates and lifts water from the sea to provide rains and rivers that flow from mountains, drives winds that move atmospheric water from the ocean to the land and cleanses the local skies of pollutants, generates soils through complex processes of forest and grassland growth and decay, and concentrates low-energy carbon from the atmosphere into higher energy tissues of a plant through photosynthesis. The use and conversion of energy amongst different forms is always governed by the laws of thermodynamics.

3.7 Energy Density

A very important component of the quality of energy is the energy density of a material, that is, the concentration of energy per unit of mass or volume. For example, the following fuels have approximately the following energy densities (MJ/Kg): diesel (48), liquid petroleum gas (46.4), gasoline (44.4), coal (24), ethanol (26), animal fat (37), sugar (17), gunpowder (3), lithium batteries (0.5), and lead acid batteries (0.2). Energy density is especially important with respect to storage: i.e., the quantity of energy that can be stored per unit of volume or mass (EIA 2013;

Wikipedia 2016). The difference between the energy density of gasoline and batteries helps explain why it is so difficult to construct inexpensive battery-driven automobiles.

3.8 Power

Power refers to the *rate* at which energy is generated or used, and is related to total energy used as water flowing from a tap at a certain rate is related to the total water caught in a bucket in a minute. For example a light bulb is rated in kilowatts, a unit of power. Power can be converted to total energy used by multiplying by the duration of use. So if a 100 W light bulb using 100 Wh (360 KJ) in 1 h, is used for 10 h the total amount of energy it uses is 1000 Wh or 1 kWh. In a day, it will use 2400 Wh or 8640 KJ—equivalent to the energy in about one quarter liter of oil. Advances in technology in the past 300 years or so have dramatically increased the capacity of humans to do work through increasingly powerful machines (Table 3.1), facilitating unprecedented economic growth in many parts of the world. As with energy, units of power are always expressed in heat equivalents— how much water would be heated by how many degrees when all of the energy in question is converted to heat. Again there is enormous confusion simply because different people or entities use different units to express power (Table 3.2). As one simple example, a 100 horsepower engine would produce 268 MJ total energy (including heat) in 1 h.

There are several additional critical things to think about when thinking about energy. First of all, there is the quantity of it, how much is either being used or is available to the species or human society using it. Second is the quality of that energy: that is, the form that it is in, which has a great deal to do with the energy's utility (Table 3.2).

Table 3.1 Evolution of power outputs of machines available to humans

Machine	Horsepower	Kilowatt	Joules per second
Man pushing a lever	0.05	0.04	40
Ox pulling a load	0.5	0.4	307
Water wheels	0.5–5	0.4–3.7	300–2800
Versailles water works (1600)	75	56	42,000
Newcomen steam engine	5.5	4.1	3070
Watt's steam engine	40	30	22,000
Marine steam engine (1850)	1000	746	55,200
Marine steam engine (1900)	8000	6000	4.5 million
Steam turbine (1940s)	300,000	224,000	165 million
Nuclear power plant (1970)	1,500,000	1,120,000	840 million

Derived from Cook (1976). Man, Energy, Society, W. H. Freeman

Table 3.2 SI units for energy and conversion from common units to SI (International System of Units) units

Getting a feel for energy units: their conversions to SI units, energy costs of materials and activities. Exponential notation: e^3 = one thousand (Kilo), e^6 one million (Mega), e^9 one billion (Giga)—exponents are multiplied by adding them, so: $1.0 \ e^3 * 3.0 \ e^6 = 3.0 \ e^9$

Useful conversions	
One calorie	4.1868 J
One Kilocalorie (Cal or Kcal)	4187 J
One BTU	1.055 kJ (Thousand Joules)
One KWh	3.6 MJ (Million Joules)
One therm	105.5 MJ
One liter of gasoline	35 MJ
One gallon of gasoline	132 MJ
One gallon of diesel	146 MJ
One gallon of ethanol	89 MJ
One cord dried good hardwood	26 GJ (Billion Joules)
One barrel of oil	6.118 GJ (Billion Joules)
One Ton of oil	41.868 GJ (=6.84 Barrels)
Some basic energy costs	
One metric ton of glass	5.3 GJ
One metric ton of steel	21.3 GJ
One metric ton of aluminum	64.9 GJ
One metric ton of cement	5.3 GJ
One MT of nitrogen fertilizer	78.2 GJ
One MT of phosphorus fertilizer	17.5 GJ
One MT of potassium fertilizer	13.8 GJ
1 J	Picking up a newspaper
1 Million Joules (1 MJ)	A person working hard for 3 h
3 Million Joules (3 MJ)	A person working hard for one day
11 Million Joules (11 MJ)	Food energy requirement for one person for one day
1 Billion Joules (1 GJ)	Energy in 7 gallons of gasoline
1 Trillion Joules (1 TJ)	Rocket launch
100×10^{18} Joules (100 ExaJ)	Energy used by US in 1 year (2009)
488×10^{18} Joules (488 ExaJ)	Energy used by world in 1 year (2005)

Thanks in part to R. L. Jaffe and W. Taylor Energy info card, Physics of energy 8.21, Massachusetts Institute of Technology.

Summary

Thus we see that energy is essential, ubiquitous, pervasive, somewhat hard to define and in general both simple and elusive. We will use the word "energy" in the rest of this book as simply "energy" as if we understood it well, or by saying that we certainly understand and can measure its effects. Energy's role in our lives would be much easier to understand and less mysterious if we were clear about how and

3.8 Power

where it is used, and if we applied a single unit of measurement (Joules) to all energy (Table 3.2). You might think this sounds impossibly difficult. How could we ever have a sense of the joules of energy required to run our computer, eat food at a Café, build a house, or fly on holiday to an exotic location? Well it can easily be done. Today we use money as our universal measure, but the value and utility of money varies over time. We may need to start thinking of money as simply our means of keeping track of energy, as a lien on energy. Meanwhile when we think about everything going on in our lives, we can think about energy use and energy both in terms of quantity and quality. All of these properties come into play when we think about investments.

References

Hamilton JD (2009) Causes and Consequences of the Oil Shock of 2007–08. Brookings Papers on Economic Activity, Spring.
Murphy DJ, Hall CAS (2011) Energy return on investment, peak oil, and the end of economic growth. Annals of the New York Academy of Sciences 1219: 52–72.
Rangel-Hernandez VH, Damian-Ascencio C, Belman-Flores JM, Zaleta-Aguilar A (2016) Assessing the Exergy Costs of a 332-MW Pulverized Coal-Fired Boiler. Entropy 18(8):300.
Run Like the Wind: A Geographical Look at Kenyan Supremacy in Long Distance running. 123HelpMe.com. 25 Jul 2016. http://www.123HelpMe.com/view.asp?id=84555.
Wikipedia. (2016) Energy Density. https://en.wikipedia.org/wiki/Energy_density.
U.S. Energy Information Agency. (2013) Few trsansportation fuels surpass the energy densities of gasoline and idiesel. http://www.eia.gov/todayinenergy/detail.php?id=9991.

Chapter 4
Energy and the Physical World

We have defined energy as that which is responsible for transformations of manner, or, less ideally, for any motion. Since the world we live in is mostly dynamic, not static, this means that energy is in and being used by essentially everything around us: the wind blowing clouds and trees, your pets or friends or family around you, all of the appliances and micro appliances that surround even the modestly well off. People are attracted by powerful motion: by windstorms, cars, athletes, dance and sports on TV, the grace and power of wild animals. In other words we are attracted to high concentrations of energy, and often want to engage them, even if we cannot really understand why. And of course we are dependent on energy in various ways, as food, transport to school or a job, recreation, and for a refreshing change of scene.

4.1 Main Sources of Energy on Earth

Three main sources of energy exist on Earth: solar, geophysical, and nuclear. In a sense nuclear is the most basic, for fusion processes operate the Sun and fission heats the Earth's interior. Nuclear and geophysical working together define the heating of the Earth's interior, which pushes the continents apart in "continental drift" and causes "hot spots" where there are volcanoes and earthquakes. Africa fits nicely against South America, as was obvious to me looking at a world map in the 7th grade, and indeed it was once lodged there. Huge sources of radioactively- formed heat cause upward movements of the Earth's magma (hot rock), often in the middle of the oceans, and push the adjacent lithosphere ("plates") away from the heat source (Fig. 4.1). Iceland with its many volcanoes is the Northern limit of the mid-Atlantic rift. The movements of the plates cause different pieces of the Earth to move and often smash into each other, generating earthquakes and volcanoes along the way. The results can be most impressive: the Himalayan Mountains, highest on Earth, were formed when India sailed across the equator and smashed into the rest of the Asian Continent. California was formed by at least three land masses crashing into

Continental Drift Diagram

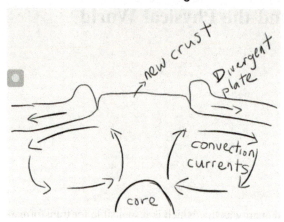

Fig. 4.1 Basic pattern of the geophysics of continental drift. "Hot spots" of high radioactive activity occur in various spots and lines deep in the Earth (called core in this diagram), such as in the middle of ocean basins. The thermal energy (heat) produced moves upward and pushes old crust aside, until it ultimately goes down into the Earth in subduction zones. Apparently the material then moves at depth back toward the source

what is now Nevada. If the plate moves downward, back into the Earth, it creates a "subduction zone" and this is often associated with volcanoes and earthquakes. The Pacific Ocean is surrounded by a "ring of fire" due to these collisions and subductions, which includes the frequent earthquakes and volcanoes, alive or dormant, that are found along the West Coasts of the Southern Andes Mountains and the United States, including Alaska, the Aleutians, Kamchatka in Eastern Russia, much of Japan, New Guinea and New Zealand. All of this was started by hot spots in the middle of the Pacific Ocean, fueled initially by radioactive activity deep in the Earth. Sometimes plates drift over a hot spot, generating an Island chain. Hawaii is one such example, with the northwestern most islands the oldest and the big island the newest, with active volcanoes on its southern shoulder.

The second main source of energy is another type of geophysical, which includes the inertia remaining from the formation of the solar system, but also the tides generated from this movement (seen mostly in ocean tides, but also in slight deformations of the Earth's crust such as the "bulge" at the equator) and also the crustal movements mentioned above. The most obvious inertial components include the daily rotation of the Earth about its own axis and the annual revolving of the Earth about the Sun. We get seasons because the Earth's axis is tilted at 23.5° relative to the plane of its revolution about the Sun, so in summer in the respective hemispheres the poles are pointed much more directly at the Sun and hence each square meter gets a larger photon flux or input of energy. Because the Earth is slowing down due to tidal friction (at milliseconds per century, so don't worry), like a slowing top, it is wobbling slightly. These wobbles, as well as other cyclical shifts in the relation of the Earth to the Sun, occur at three distinct thousands of year

4.1 Main Sources of Energy on Earth

cycles, are called Milankovitch cycles and are thought the principal cause of long term climate cycles, such as ice ages. The Earth has had something like 12 distinct ice ages in its long history, and the relative warmth we enjoy now is not the normal situation over geological time.

The Sun is the source of energy for most of what we see from day to day on Earth. It warms the atmosphere, operates the hydrological cycle and allows and feeds life. Eventually essentially all of this high-quality input of photons is transferred—at each work step and each transformation—to low grade heat in accordance to the second law of thermodynamics. The heat is then reradiated back to space as low-grade heat, maintaining the Earth at a relatively constant temperature. When the Sun's energy strikes the Earth's surface that portion that is not reflected does many types of work on the Earth's surface. We can feel the effects in the heating of dark surfaces, but the largest amount of work that sunlight does on Earth is to evaporate water. This is because it takes a great deal of energy to transform water from a liquid to a gaseous phase (the latent heat mentioned in Chap. 2). Wind and more generally weather is caused by the uneven heating of the Earth's surface by the Sun. At a larger scale, the Sun heats the Earth more at the equator than toward the poles because the incident radiation is more intense there because the land is perpendicular to the photon flux. You can see this for yourself by holding your hand at an angle to the Sun's rays on a bright sunny day and then slowly moving your hand so that it is perpendicular to the rays. You can feel your hand get warmer because more photons are striking each square centimeter of your hand. Similarly more photons strike the Earth in the vicinity of the equator, so that the equator tends to warm more than other portions of the Earth. In general this heat tends to be moved North and South by the oceanic (e.g., Gulf Stream) and especially atmospheric systems. There is a very special mechanism and pattern for this which we examine next.

4.2 Hadley Cells

Early sailors of the great oceans had a tough time of it. Square-rigged sailing boats had great difficulty in going into the wind, especially before the invention of effective keels and rudders. Fortunately for the sailors and those who invested in their voyages it did not take them too long to figure out that while the wind could blow from any direction at any time, there were some clear patterns that on average made their voyages much, much easier. For example the Portuguese were the first to sail from Europe around the tip of Africa to the lucrative spice trades of the orient. But to go around Africa while staying in sight of land requires sailing almost directly into the wind, as does returning when near to Portugal. Captains under Prince Henry the Navigator figured out that if the ships sailed out into the ocean, for example starting Southwest instead of East when going around Africa they could then catch much more favorable winds and shorten the journey by many weeks (Fig. 4.2). In tropical regions, between 30° North and South, the winds tended to blow regularly from East to West. These winds were later called "trade winds" for obvious reasons. But in temperate regions, North and South of the tropics, winds (Westerlies) tended to blow from West to East. Knowledgeable European sailors going from Europe to North America would use the trade winds for moving from Europe to the Americas and then return further North, to use the Westerlies to return home, while avoiding, as much as possible, the doldrums on the equator and the horse latitudes at 30° where air masses move vertically rather than horizontally. The Pilgrim ship Mayflower, iconic in American history, took 66 miserable days to go from England to America sailing into the Westerlies, but fewer than half that to come back.

It took a while to figure out why the winds blew in the patterns they did. British Meteorologist George Hadley was the first to offer an explanation, while figuring out the first (equatorial) cell in 1735 that bears his name. One result of the intense heating of the Earth's surface on the equator is that there are very strong upward movements of air masses there as the warmth of the Earth heats the overlying air, and that warm air rises because it is less dense. Because the Earth's surface and the air itself are warm that air tends to hold a lot of water. You can think of it this way: when air molecules are warm they move more rapidly, and this rapid motion keeps hitting and bouncing the water molecules, keeping them in suspension. As the air masses rise, however, they cool and generate clouds, which are readily seen in satellite pictures of the Earth as a band of clouds in the vicinity of the equator. When the air masses have reached some four or five miles (10 km) of altitude they cannot rise any further because they have expanded and cooled, although additional air keeps piling up underneath them. This continual "piling up" of air at the equator creates a region of high pressure near the top of the atmosphere which pushes the air masses North and South, so there are huge currents of air that move North and South from the equator at about 5 miles height. When these air masses reach about 30° North and South they have cooled enough so that they are now denser than the air below them, and they descend. As they descend they generate a high pressure mass of air at the surface of the Earth at about 30° North and 30° South, which in turn pushes air masses at each location both North and South. In the Northern hemisphere this generates winds that flow to the low pressure area at ground level near the equator, thus closing the loop.

An additional factor is the Coriolis effect, caused by the rotation of the Earth. As the surface air masses travel from 30° toward the equator they maintain their inertia but incrementally enter areas where the Earth's surface is spinning more rapidly, as there is a larger circumference to rotate over 24 h. This gives the winds the appearance of being shifted to the right in the Northern hemisphere and to the left in the Southern hemisphere (Fig. 4.2). The net result is the very steady trade winds of the tropics. A second loop, also driven by the high pressures of 30° North and South, cycles from about 30° to 60°, creating (with the help of the Coriolis effect) the Westerlies that are familiar to those living in the temperate regions as they watch storm systems move across the land from west to east (Fig. 4.3).

4.2 Hadley Cells

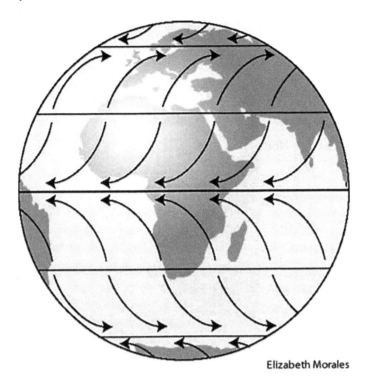

Fig. 4.2 Basic wind patterns on the Earth's surface

The Sun also drives more local weather patterns, principally by unequal heating of a region. Dark areas, such as plowed fields or coniferous forests, absorb more solar energy which then heats the atmosphere above them and causes the air to rise, which in turn causes a low pressure region. Other areas, less heated, will have higher pressure and the air masses will flow from high pressure to low, which we experience as breezes and winds.

For humans, and indeed life generally, the most important work, besides photosynthesis, that the Sun does on the Earth is to operate the hydrological cycle, evaporating and purifying water and lifting it high in the atmosphere and to mountains where it falls as rain and snow that causes rivers to flow. Lemon et al. (1971) measured all the work that the Sun did in a corn field and determined that evaporating water either directly or through plant leaves (called transpiration) was quantitatively the most important.

Finally the Sun does all kinds of other work (see e.g., Miller 1981). Its ultraviolet rays are powerful bactericides, and so laundry left on the clothesline in the Sun will be in fact cleaner than if not left in the Sun. But these ultraviolet rays are very

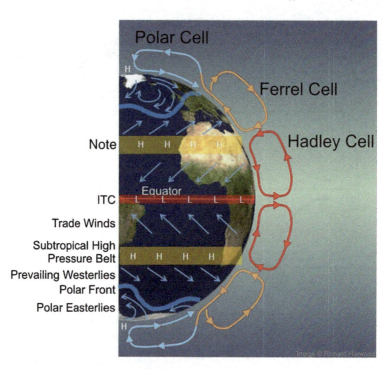

Fig. 4.3 Basic pattern of Hadley cells. The Earth rotates from *left* to *right* in this diagram, causing North and South moving winds to be shifted to *right* in Northern Hemisphere and *left* in Southern Hemisphere. *Note* similarity of Hadley cells (near equator) to the cycles of the lithosphere in Fig. 4.1, with a "hot spot" (equator in this case) acting as a heat source driving the bidirectional circulation cycles. *Source* Richard Harwood meteorology notes

potent to all life, and can cause damage to phytoplankton, to the eyes of deer and other ungulates, to our skin and maybe much more of life. And of course solar energy runs photovoltaic systems.

4.3 "Investments" by Nature

While it is difficult to think of inanimate nature making investments—things just are, many past cosmological, geological, and evolutionary accidents and designs have occurred on the planet that have heavily benefited humankind. In a way the Earth has made enormous investments on behalf, so to speak, of mankind. First, is the amazing occurrence of Earth itself at just the right distance from the Sun, with lots of water and ecosystems that regulate the atmosphere, made the oil and generated a milieu within which humans and so many other species can live and thrive. For many this is evidence for a god that has concern about humans. But modern astronomers and cosmologists have found in the mind numbing number of galaxies and solar systems "out there" probabilities for thousands of such planets, many of which would seem to have the characteristics needed for life. Whatever your perspective is, we must all be amazed at how the Earth is such a hospitable home for humans and their endeavors.

Of particular interest to this book is how the Earth, over a 100 million and more years, has generated a vast storehouse of concentrated solar energy in the form of fossil fuels (Dukes 2003). These fuels took untold immense amounts of ancient sunlight, captured by earlier ecosystems, and many tens to hundreds of millions of years of intense geological energies to generate what are, from the perspective of humans, nearly perfect fuels, which we will burn in just a few centuries. Fossil fuels come out of the ground almost ready to use. Coal and natural gas can generally be burned as is after only a little cleaning. Oil needs to be refined a bit before it can be used.

4.4 The Formation of Fossil Fuels

Very special circumstances were required for the formation of fossil fuels that are so important to our modern life. Coal, oil, and gas are organic materials, that is, they are plant and animal remains composed mostly of reduced (i.e., hydrogen rich) carbon as is all life. As plant life evolved some 3 billion years ago a great deal of organic material was formed, most of which was oxidized relatively soon and turned back to carbon dioxide in the atmosphere, which is not an energy source but is available for new plant growth. Some very small part of this organic material found its way to *anaerobic* (meaning without oxygen) basins, such as deep lakes or marine areas for oil and gas, and hence accumulated as various deposits. Chemically natural gas (methane) has four molecules of hydrogen per molecule of carbon, oil has about equal amounts, and coal is mostly carbon, although with small amounts of hydrogen and sulfur and trace amounts of many elements, including troublesome mercury and uranium. Thus there is a progressive increase in CO_2 per unit of energy delivered from natural gas to oil to coal, with coal releasing almost double the CO_2 per heat unit relative to gas.

The creation of exploitable oil and gas fields has been quite rare in the geologic past. It happened mostly some 90 and 150 million years ago when the Earth was very warm, and in very special and limited environments (Fig. 4.4). The time required to turn the organic source material into oil and gas is extremely long and requires the organic material being buried at just the right depth (about 3000 m or two miles) and temperature (about 100 °C) to "pressure cook" the organic material into oil. As a consequence, significant quantities of commercially exploitable oil and gas are found in only a relatively few regions of the Earth's surface. Coal, formed in great fresh water swamps, required far less stringent conditions for its

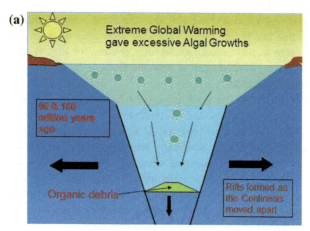

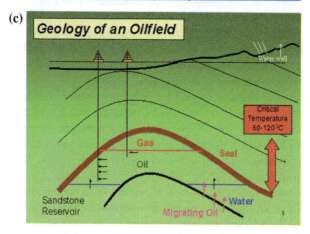

Fig. 4.4 The usual formation of oil and gas. **a** A deep trench is generated in the Earth from crustal movements, such as when Scotland drifted away from Norway. Marine life including phytoplankton and zooplankton grow in the trench and fall to the bottom. If the trench is deep and warm enough anaerobic (no oxygen) conditions exist and the organic material accumulates for many millions of years. **b** If this is followed by heavy rains sediments fall onto the organic material, further protecting it from oxidation and "pressure cooking" it into shorter length carbon chains. An ideal depth for this is about 3000–4000 m. **c** The light oil and gas move upward through the Earth and most is lost to the atmosphere. Some small part is trapped by impermeable geological structures, such as anticlines of sandstone, and remains as oil and gas deposits, which can then be exploited (Images courtesy of Colin Campbell)

production and is more common. Gas too is widely dispersed, but large reservoirs are relatively rare. On the other hand gas is found widely at low concentrations associated with coal and in "tight" shales and sandstones. Exploitation of these diffuse resources is becoming increasingly important as the large gas fields found earlier face serious depletion. Whether or not these newer "unconventional" fields can maintain production at the present level for very long is unknown at this time (Miller and Sorrell 2014).

4.5 Oil

Petroleum (from Greek "rock oil") is sometimes used to mean just oil, but technically petroleum means oil and gas and some semi-solid components. "*Conventional*" petroleum means oil and gas derived from geologic deposits, usually found and exploited using drill bit technology with the resources moving to the surface because of their own pressure or with additional pressure supplied by pumping additional natural gas or water into the reservoir. "*Unconventional*" petroleum includes shale oil, tar sands, some bitumens, and coal-bed methane. These resources are often "undercooked" (oil sands) or "overcooked" (tar sands) by geological forces, and/or of a less concentrated nature (for example mixed with sand) or found in deposits in very deep or hostile environments. A somewhat gray area is "previously uneconomic" oil and gas which is usually a lower grade resource of any kind that was traditionally not worth exploiting, but becomes commercially exploitable when oil prices are high, with "new" technologies such as horizontal drilling, and with the depletion of easier to exploit and cheaper traditional reservoirs. Essentially all unconventional oil requires more energy to extract.

Humans have tended to exploit the large, high quality and easy oil deposits first. This is an example of the "best first" principle introduced by the Nineteenth century economist David Ricardo. Today about half of the oil we extract comes from only three percent of the oil fields, most of which have been exploited for 50 or more years (Simmons 2002). Onshore deposits in places such as Texas and Louisiana were developed long before deeper offshore regions. But the large onshore resources are now depleted, and there are more than 4000 very expensive platforms in the Gulf of Mexico off Louisiana and the mouth of the Mississippi River that are responsible for much of the United States' remaining oil and gas production. Exploitation of the North Sea deposits has involved similar, expensive technology. Over time we are moving further and further off shore, and deeper and deeper in the sediment, and are finding on average smaller fields. As of this writing arctic exploration and development has been postponed because of the large expense and disappointing results so far, combined with the fall in the price of oil since mid-2014.

4.6 Natural Gas

"Petroleum" includes natural gas liquids and natural gas. Natural gas is often found associated with oil, although it has other possible sources, including coal beds and organic-rich shales. We get natural gas when the original plant material, with molecules often of hundreds to thousands of carbon atoms linked together, has been "cracked" or broken by geological energies to a length of five or fewer carbon atoms, becoming most usually methane, with one carbon atom surrounded by four hydrogen molecules. Once distribution systems are built, gas is an ideal fuel as it is easy to handle and very useful. Since oxidizing hydrogen releases more energy per unit carbon dioxide produced, its use would contribute less to climate change than other fossil fuels, at least if not much of it leaks from the wells and pipes (methane has a global warming potential of 28–84 times higher than carbon dioxide per molecule; Myhre et al. 2013; Howarth 2015). When natural gas is held in a tank some heavier fractions fall out as natural gas liquids, and these materials can be used either directly or as inputs to refineries. While natural gas was once considered an undesirable and dangerous byproduct of oil production and flared into the atmosphere, with time a complex pipeline system evolved and now natural gas more or less ties with coal as the second most important fuel in the United States and the world. An important question is: if oil supplies falter can natural gas take over its role? While it is not as energy dense or transportable as oil it comes close, and because it is relatively clean and malleable it has many special uses such as a fuel for baking and as a feedstock for plastics and nitrogen fertilizer. Natural gas is used increasingly to make "clean" electricity, although it is possible that our grandchildren will be unhappy about this if they need that gas for higher value uses such as making fertilizer.

4.7 Unconventional Petroleum: The Fracking Revolution

There is a great deal of excitement and debate as of the second decade of this century about whether "unconventional" (but light, good quality) oil from, for example, the Bakken formation in North Dakota and the Eagle Ford field in Texas,

4.7 Unconventional Petroleum: The Fracking Revolution

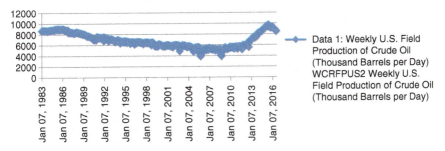

Fig. 4.5 Time series of oil production in the United States 1983–2016 The peak in US oil production of 10,500 thousand barrels per day occurred in 1970 (*Source* U.S. EIA; http://www.eia.gov/dnav/pet/hist/LeafHandler.ashx?n=PET&s=WCRFPUS2&f=W)

and natural gas from shales such as the Marcellus shale can continue to provide an energy renaissance for the United States. Production of oil in the U.S., which had been falling since 1970, increased rapidly from 2008 to 2014 and almost reached a new record. Since mid-2015 though, with oil prices falling and the best areas heavily exploited, oil production has been falling again (Fig. 4.5). While the amount of oil in these formations is enormous, the rocks (actually sandstone, not shale) have low porosity (pore space) and permeability (ability of oil to flow through the formation) so that only some 5 % of the oil in place can be extracted even by new, energy intensive "heroic" efforts. This compares with an average of some 38 % from conventional fields (Duffeyes 2001). The new technologies include horizontal drilling (with up to 2 mile lateral extensions) and the shattering or "fracking" of the rocks with very high-pressure water and chemicals. This has allowed considerable amounts of previously inaccessible oil and gas to be produced (Fig. 4.6). Optimists predict one hundred years of gas and that the US will soon export oil (as of 2016 it imports half of its use). But so far most of this "unconventional" oil and gas has come from a relatively few "sweet spots," so the total production may go through nearly a full exploitation cycle in just a few decades (Hughes 2015). When large claims are made, as for Monterrey oil, it is always good to have a competent geologist such as Hughes take a look. Meanwhile most estimates for total global oil and gas production suggest peaks and then declines within a decade or two, with or without unconventional sources (Fig. 4.7).

Meanwhile conventional gas production in the United States has peaked and declined to less than half the peak, so that so far the unconventional gas of all kinds is mostly compensating for this decline. Where it falls short, imported gas bridges the gap. Thus while fracked oil and natural gas is likely to be very important as conventional oil production and availability declines, it is likely to extend the petroleum age by only a few decades. As of this writing (August 2016) both unconventional and total oil production in the United States have already gone through a secondary peak and are declining (U.S. EIA).

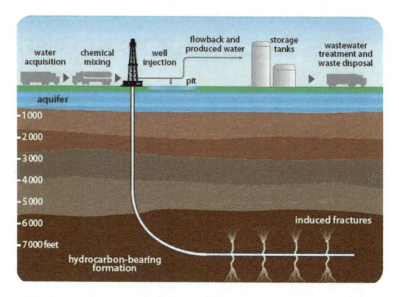

Fig. 4.6 Horizontal drilling and "Fracking" technology. As a hole is drilled to the depth of the oil bearing formation, which is often source rock, then the drill is turned sidewise and drilling continued for up to 4 km. Then very high pressure water with chemicals and sand is pumped into the hole to fracture the rock and release the oil held tightly there

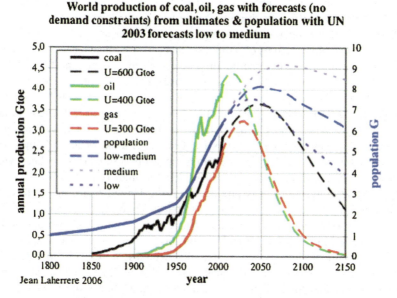

Fig. 4.7 Oil, gas, and coal production and use for the world, and projections based on supply constraints. Similar projections (although with a sharper decline in coal) can be found in Mohr et al. (2015) and Maggio and Cachiola (2012). Gtoe means GigaTons oil equivalent (as energy). Graph courtesy of Jean Laherrre

"Heavy" (i.e., long chain, high molecular weight) oil tends to be relatively "overcooked" compared to conventional "light" oil, is very abundant in Canada and Venezuela, and can be exploited by digging out the heavy oil or the "tar sands" and treating it with natural gas to add hydrogen and make the oil more fluid. This is, in general, an inefficient and capital—and energy—intensive process. Nevertheless the technology to exploit these deposits exists, the resources are large (but again of varying quality) and are likely to continue to supply a modest quantity of petroleum to the world. The exploitation of Canadian tar sands is especially well developed, but their extraction has been declining since 2014 with the decline in the world price of oil.

4.8 Coal

Coal was the backbone of the initial industrialization of the United Kingdom and of Germany, starting around 1700, and the United States a century later. It was the most extensively used fossil fuel until surpassed by oil in the middle of the last century. While coal is much more widely distributed over the Earth's surface than oil, with especially large reserves in Russia, the United States, Australia, and China, the depletion of some of the best resources was a matter of grave concern as early as the mid-Nineteenth century in Britain (Jevons 1865). Estimates of the resource base of coal seem less volatile than for oil or gas. Nevertheless recent assessments by Maggio and Cacciola (2012) and also Mohr et al. (2015) give a most probable estimate for global coal of some 1200 billion metric tonnes, with the largest component being medium quality bituminous. Earlier estimates of how long coal would remain a viable resource, based on dividing the resource base by the annual use, were from 100 to 400 years. More recent estimates of Mohr et al. and Maggio and Cachiola are for a continued sharp increase in coal use, mostly driven by the large increase in Asian coal consumption, a peak by 2025 (or 2050 in the high resource scenario) and then decline, again led by China. A recent paper by Tang et al. (2016) indicate that Chinese coal consumption has already peaked in 2014. The reason that the newer estimates for how long coal will last are based on lower estimates of total reserves, and, especially, the lower probability given for exploiting deep thin seams.

The future of coal (and oil and gas) as a resource may be driven more by global efforts to decrease CO_2 emissions than by continued economic growth and the price of this relatively carbon—intensive fuel. Most predictions of energy use (e.g., US EIA) have coal consumption continuing at present rates or increasing. On the other hand, higher costs of all fuels along with a reduction in economic growth in many areas may generate a drive toward lower and/or more efficient use of fuels and cause a decline in energy use more quickly than once thought. Only one thing is clear: the future of fossil energy use is highly uncertain.

References

Duffeyes K (2001) Hubbert's peak: the impending world oil shortage. Princeton University Press, Princeton.
Dukes JS (2003) Burning buried sunshine: human consumption of ancient solar energy. Climatic Change 61:31–44.
Howarth R (2015) Methane emissions and climatic warming risk from hydraulic fracturing and shale gas development: implications for policy. Energy and Emission Control Technologies. 3:45–54.
Hughes D (2015) Shale gas reality check: Revisiting the U.S. Department of Energy Play-by-Play Forecasts through 2040 from Annual Energy Outlook. Post Carbon Institute.
Jevons WS (1865) In: The coal question: an inquiry concerning the progress of the nation, and the probable exhaustion of our coal mines (ed) Flux AW, 1965. Reprints of Economic Classics. Augustus M. Kelley, New York.
Lemon E, Stewart W, Shawcroft EW (1971) The sun's work in a corn field. Science 174:371–378.
Maggio G, Cacciola G (2012) When will oil, natural gas, and coal peak? Fuel 31(98):111–123.
Mohr SH, Wang J, Ellem G, Ward J, Giurco D (2015) Projection of world fossil fuels by country. Fuel 1(141):120–135.
Miller DH (1981) Energy at the surface of the Earth. An introduction to the energetics of ecosystems: Academic Press, New York.
Miller RG, Sorrell SR (2014) The future of oil supply. Phil Trans R Soc A 372.
Myhre G, Shindell D, Bréon F-M, Collins W, Fuglestvedt J, Huang J, Koch D, Lamarque J-F, Lee D, Mendoza B, Nakajima T, Robock A, Stephens G, Takemura T, Zhang H (2013) Anthropogenic and natural radiative forcing. In Climate Change 2013: The Physical Science Basis. Contribution of Working Group I to the Fifth Assessment Report of the Intergovernmental Panel on Climate Change (ed) Stocker TF, Qin D, Plattner G-K, Tignor M, Allen SK, Doschung J, Nauels A, Xia Y, Bex V, Midgley PM, 659–740. Cambridge University Press.
Simmons M (2002) The World's giant oil fields. Simmons and Company.
U.S. Energy Information Agency: http://www.eia.gov/dnav/pet/pet_sum_sndw_dcus_nus_4.htm.
Tang X, Xu T, Jin Yi, McLellan BC, Wang J, Li S et al. (2016) China's coal consumption declining—Impermanent or permanent? Resour Conserv Recy. doi:10.1016/j.resconrec.2016.07.018.

Part II
Energy and Biology

The development of the basic concepts of thermodynamics and energy more generally were undertaken within the fields of physics and chemistry. But it did not take long for many great thinkers, such as the physicist Ludwig Boltzmann, to understand that the priciples of energy also applied to biology, that in fact energy also underlay, and in fact determined, nearly everything pertaining to life. The next three chapters discusses how biology, including physiology, ecology and evolution, is all about energy, investments and EROI.

Part II
Energy and Biology

Chapter 5
The Ecological Theater and the Evolutionary Play

Humans as biological organisms must operate within the basic laws of thermodynamics. But starting with Ludwig Boltzmann in the 1880s, many great thinkers have delved more to deeply, to uncover how life itself, and all of biology, is essentially about obtaining and using energy. In other words, all of life is about making energy investments to gain more energy with the expectation of continuing their own life and sending genes into the future. To this end, I propose the *iron law of evolution*: organisms must extract more energy by exploiting their environment than they expend in doing so. This expenditure includes the energy costs of maintenance metabolism, adaptation to environmental exigencies, resource exploitation, and reproduction.

As I write this, I am planting my garden for the summer. I look at seeds and think how each is a small globe of energy gathered and invested by their parents, together with a genetic information packet, with the expectation that this investment will carry the parent's genes into the future. Likewise the seed invests in roots and the first few leaves, with the expectation that this investment will generate a large enough return to allow the little plant to grow, add more leaves, and so on. Usually it works, although with cultivars there is the intervention of the seed companies to modify nature's original plans! Meanwhile around me all of nature, from the Douglas firs and Ponderosa pines of our yard to the swallows and robins, are similarly investing in capturing more energy and using it for maintenance metabolism, growth, reproduction, and nurturing young. Similarly I invest my own energy into tilling the soil, weeding, irrigating, and so on.

This perspective of energy investment, although not widely thought about or taught, had great influence on many of our most thoughtful and influential scientists including in the twentieth century Alfred Lotka, G. Evelyn Hutchinson, Leigh Van Valen and Thomas and Eugene Odum. They have endeavored to explore this perspective by synthesizing energy with ecology and evolution to examine how, in overview, the biological world works.

The ecological-evolutionary concept was laid out elegantly and even poetically by G. Evelyn Hutchinson in the title of his 1965 book *The Ecological Theater and the Evolutionary Play*, which emphasized how the evolution of life (the play) has played out in the ecological setting (the theatre) of constraining and changing environments. The environment supplies the energy and materials to produce and maintain living organisms and determines where and in what forms life can exist. It also offers the prerequisite energy and materials that allows survival and reproduction to move forward.

5.1 Energy and Biology

The great geneticst Theodosius Dobzhansky (1973) famously wrote "Nothing in biology makes sense except in the light of evolution." We can paraphrase this statement as: "Nothing in biology makes sense except in the light of ecology, and nothing in either ecology or evolution makes sense except in the light of energy." The laws of thermodynamics and conservation of energy, matter, and chemical stoichiometry (essentially elemental balance) are as relevant to ecology and evolution as they are to physics, engineering, and molecular biology. But there are many additional aspects that must be considered when we look at the large complexity and diversity of life (Fig. 5.1).

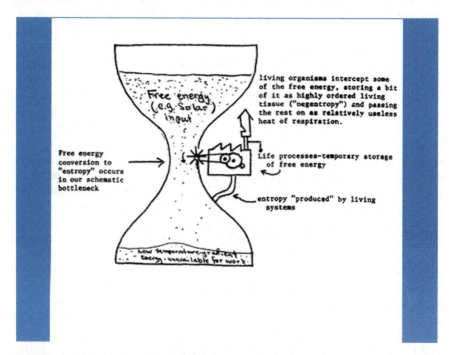

Fig. 5.1 A mechanical analogy to the thermodynamic processes by which life gains energy for its own activities. The hourglass represents the continuing passage of free energy (i.e., the Sun's output) to unavailable energy. The paddle wheel represents living creatures' interception of some of this energy to maintain otherwise unlikely biotic structure

5.1 Energy and Biology

Most of the energy available to organisms on the Earth became available originally in the form of the photon flux from the Sun. Sometimes we think we can see sunbeams, but what we actually see is the effect of photons on dust in the air. The Sun continuously consumes itself in the process of nuclear fusion, in the process emitting a huge number of photons to space, some in the direction of Earth. Approximately 5.7×10^{24} J of solar energy is irradiated to the Earth's surface on an annual basis. A very small part of this energy, roughly 0.1 % or some 3×10^{21} J/year, is captured by green plants in the chlorophyll molecule (Figs. 5.2 and 5.3). In this process of *photosynthesis* energy from the Sun is captured by green plants using chlorophyll, a very special compound similar in structure to the hemoglobin in our own blood. Chlorophyll, and plants more generally, appear green to our eyes because they absorb the shorter (and more energetic) blue and longer (and less energetic) red wavelengths from the Sun and reflect back the green wavelengths that they do not use.

A thick layer of green plants cover the Earth wherever temperatures are moderate and water is abundant. The amount of energy captured by plants during daylight in the growing season supplies the energy to run all of life: plants in day and night, animals and microbes for all hours of the year. The first step in energy capture occurs in the center of the chlorophyll molecule where electrons circling the magnesium–nitrogen complex in the center of the molecule are "hit" by a photon from the Sun and energized by being "pushed" into a larger, higher energy orbit. This allows them to store more energy and then pass it to special chemical compounds. Think of an ice skater spinning with her arms extended. If her partner gives her a well-aimed push, she can spin even faster with her arms out in a more energy-intense state. When she brings her arms in close to her torso, her spinning is accelerated but her total energy is the same.

When in photosynthesis the electron drops back into a lower orbit, the energy in that electron is captured by various biochemical pathways, and ultimately as reduced carbohydrates or hydrocarbons. The initial electron donor is a chemical

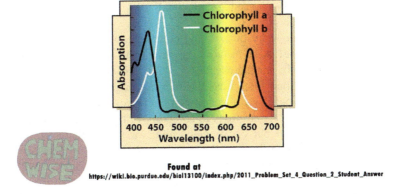

Fig. 5.2 Absorption of light by the chlorophyll molecule. The *green* wavelengths are not absorbed, hence plants look *green* to us

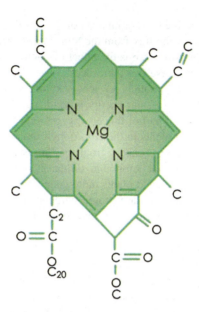

Fig. 5.3 The chemical structure of chlorophyll. The Mg and N atoms share electrons that can capture energy from photons by being "bumped" into an outer shared orbit

species excited by the absorption of light. That energy is then stored temporarily in reduced compounds such as NADP, which are then used to split water to get hydrogen and electrons.

$$2H_2O \rightarrow 4H^+ + O_2 + 4e^-$$

Once water is split, the electrons, along with H^+ from water, are combined with the carbon in CO_2, reducing it to sugar and releasing more oxygen. Free electrons are normally made available from such *reduced* compounds as sugar (which are essentially hydrogen rich, with an energized electron). The electron is passed to a series of electron acceptors (meaning compounds with a greater affinity for electrons), doing work in the process.

Then all the work done by other living organisms is undertaken, directly or indirectly, by the flow of these electrons in oxidation—reduction reactions (e.g. Lehninger et al. 2008). In nonphotosynthetic organisms, the source of electrons is reduced compounds in their food. Electrons move through various complex metabolic intermediates to specialized electron carriers in enzyme-catalyzed reactions. Those carriers in turn donate electrons to acceptors with higher electron affinities, ultimately usually oxygen but occasionally sulfur or some other element, with energy released at each step. The useful energy is stored in tiny chemical "batteries" called ATP, (adenosine triphosphate) and released in the "motors" of life, the mitochondria. This energy then is released and the electron passed to oxygen when an organism does the work of life: moving chemicals against gradients, maintaining biological structures against the forces of entropy, causing muscles to contract, and

5.1 Energy and Biology

so on. All of these are energy investments to allow the organism to further exploit environmental resources. CO_2 and water are the waste products of plant and animal respiration and represents the "ground state" (i.e., minimum energy state) of this process. The electrons are reenergized only when green plants give a new kick to the electrons when a photon from the Sun again drives photosynthesis. And hence the process continues, with the energy from solar-derived photons driving every biological activity including the movement of my fingertips on my computer's keyboard as I write these lines. It is incredible!

Something that is normally missed in discussion of either electrical or biological energy is that the *"redox"* (meaning reduction/oxidation, or electron transfer) energy transfer processes are very similar and rather general. Electrical circuits operate by a battery or generator giving electrons an energy kick (similar to photons giving electrons a kick in the chlorophyll molecule), sending them around a copper wire circuit, meanwhile doing work, to an oxygen-like ground, in an analogous fashion. The source of electrons in an electrical circuit is a battery containing two chemical species that differ in affinity for electrons (or a generator). The electrical wires provide a pathway for electron flow from the chemical species at one pole of the battery, through a motor, to the chemical species at the other pole of the battery. Because the two chemical species differ in their affinity for electrons, electrons flow through the circuit, driven by a force proportional to the difference in electron affinity of the two elements, also called the electromotive force. The electromotive force (typically a few volts) can accomplish work if an appropriate energy transducer such as a motor is placed in the circuit. The motor can be coupled to a variety of mechanical devices to accomplish work.

Likewise in food chains the high-energy, reduced compounds found in plant tissues are eaten by herbivores and then carnivores, passing the energized electrons in chemical bonds from one trophic level to the next throughout the food chain. Thus although we cannot see it, we are able to use the energy in a hamburger by oxidizing the reduced matter in the plant and animal tissue. This energy initially was obtained by the cow when it ate grass that had in turn captured that energy from the photons, and then passed it as chemical bonds to the cow and then to us. Even when we are driving a car, we are oxidizing formerly reduced plant material (oil) that is constructed of high-energy chemical bonds originally made with energy captured from the Sun by algae.

5.2 Fuels

Fuels tend to be chemically reduced compounds. That means they are hydrogen rich and oxygen poor, so that a fuel is generally a hydrocarbon, like oil or occasionally a carbohydrate such as alcohol (the "ate" on the end refers to the presence of oxygen, so that a carbohydrate will have somewhat less energy than a hydrocarbon per gram but still enough to be used as a fuel). When a reduced fuel is oxidized energy is released, along with hydrogen (as water, H_2O) and carbon (as carbon dioxide, CO_2). *Respiration* is the term for burning fuels in biological processes, and *combustion* for burning fuels in mechanical engines. The general equation for respiration or combustion of a hydrocarbon is

$$CnH_2n + O_2 \rightarrow H_2O + CO_2$$

The exact numbers required to balance the equation depend on the exact form of the hydrocarbon burned but are for oxidation of common biological foods

$$C_6H_{12}O_6 + 6O_2 \rightarrow 6CO_2 + 6H_2O$$

Like everything in the physical world, the energetics of organisms are governed by fundamental laws and principles of physics, chemistry, and biology. The most important physical–chemical principles are the laws of thermodynamics, conservation of energy and matter, and chemical stoichiometry (note that the number of each element on the left of the equation is the same as that for the right). They are well established, universal, and to our knowledge without exception. The most important biological principles are the biochemical reactions of metabolism, the laws of inheritance, the process of natural selection, and the scaling of structure and function with body size. These are also well established, universal, and apparently without exception. They apply to all organisms, from unicellular microbes to multicellular plants and animals. Organisms living today have been selected over millions of years through countless incremental combinations of these principles to acquire, transform, and allocate energy to maintain themselves as highly organized systems far from thermodynamic equilibrium, that is, far from the random state of material that they would be without the energy investment.

At all levels of biological organization, energy is transformed by metabolic processes and used to build structure and power activity. From molecules, organelles, cells, tissues, organs, individual organisms, populations, ecological communities, and the biosphere, the story is the same—energy is transformed by metabolic processes and used to build structure and power activity. Without this energy investment, the systems collapse toward equilibrium or entropy: molecules break down, cells and organisms die, populations go extinct, and ecosystems collapse. Without organisms, the Earth would be a harsh, inhospitable planet. The metabolism of living organisms has created a biosphere that contains a narrow range of physical–chemical conditions conducive to life. For the past 3.5 billion years, life has persisted, adapting to changing environments, and with tremendous diversification in form and function. The history of life is an unbroken chain of ancestors and descendants, extending back to an origin from which astronomical numbers of organisms and maybe a billion species have lived, reproduced, and died. All of these remarkable forms of life inhabiting our remarkable planet have been shaped by one universal driver: to extract energy and material resources from the environment, transform them into uniquely biological structures and functions, and do the work of survival and reproduction. The iron law of evolution is that organisms must extract more energy by exploiting their environment than their energy cost of doing so, including the costs of maintenance metabolism, adaptation to environmental exigencies, resource capture, and reproduction.

Of special interest, the evolution of atmospheric oxygen as terrestrial green plants evolved was at first an "environmental disaster" for the existing anaerobic organisms of the Earth because of the corrosive effect of oxygen. This led to the extinction of much ancient life. But it also provided a wonderful opportunity for an increase in the rate at which life could utilize foodstuffs, because respiration derived about four times more energy from carbohydrates when oxygen was available as a terminal electron acceptor than when it was not. Somewhat curiously, many of the energy pathways had already been evolved. Some of these processes worked, and although not particularly well suited to the new oxygenated environment, were retained in all of today's plant and animal cells as a sort of ghost of the Earth's past anaerobic environment (Falkowski 2006). This is one of many ways that today's life is not perfectly suited for today's environment, but rather represents a sort of hodgepodge, with selection for new processes that worked better but with the need to maintain necessary technologies that already work. Life, like many things, exists as a combination of conservative and innovative processes. This cannot be explained at all by alternative theories for the origin of life, but is well explained as a combination of natural selection and known Earth history.

5.3 Metabolism

All organisms use the metabolic reactions of respiration to break down organic compounds and capture energy to power their biological processes. They extract energy, materials, and sometimes information from the environment to maintain internal homeostasis; to move, grow, and reproduce; to encode, decode, and pass on genetic information, in other words to maintain and pass on the chemical integrity of life as energy and molecules are extracted from food and used to reconstruct molecules damaged or ruined in the struggle to maintain normal life activities. But they do so at different rates: large organisms tend to operate more slowly than small organisms (Brown et al. 2004; Sibly et al. 2012).

Metabolism is usually quantified by measuring the oxygen consumed by an organism at rest or during various activities. One of the most general characteristics of this oxygen or energy use is that smaller organisms use much more energy per gram than larger organisms. There is a similar relation between production (i.e., growth of new tissue) and size. Much of this seems to be driven by surface to volume ratios: smaller organisms have a larger ratio of more energy intensive "skin" compared to less energy intensive interiors. Small organisms use energy more intensively and use it to procure more resources from their environment which they transfer into growth. Eukaryotic organisms vary in body mass by more than 16 orders of magnitude, from microscopic single-celled algae and protists weighing less than 10^{-8} g to giant whales and trees weighing more than 10^8 g. These organisms differ by orders of magnitude in rate of metabolism and biomass production, and in longevity and generation time. This variation in metabolic and life history traits with body size is the subject of a longstanding and still growing literature on metabolic scaling. Although many details remain to be resolved, we know that the rate of mass-specific metabolism, both respiration and production of new biomass decreases with increasing body size (M) in a power relation

$$P = P_0 M^\alpha \tag{5.1}$$

Equation 5.1 can be linearized by taking the logarithms of both sides. When plotted on logarithmic axes, the scaling exponents, α and β, give the slopes, and the normalization constants, P_0 give the intercept (Fig. 5.4). Although the available data on production rates and especially on generation times are scanty, we can use the enormous range of body sizes to parameterize and obtain accurate scaling relations

$$P = 4950 M^{-0.25} \tag{5.2}$$

where P is production of new tissue in kj/day and M is in g.

Note that the scaling exponent is 0.25 and is an example of the pervasive "quarter power" allometric scaling of many biological traits, including "Kleiber's law" for scaling metabolic rate with body mass (Kleiber 1932; West et al. 1997; Sibly et al. 2012).

Thus one of the most important laws in biology is that smaller organisms live a more intense life style: they have a higher metabolic rate per gram of size, and they

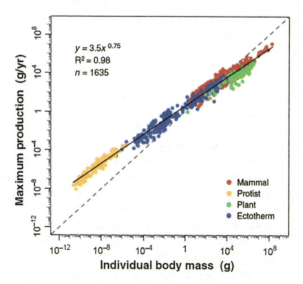

Fig. 5.4 Metabolic rates (production in this case) of organisms from the tiniest virus to blue whales (from Hatton et al. 2015). Protists are "primitive" single or few cell organisms. Ectotherm is cold blooded including invertebrates, fish, reptiles, and amphibians

also have a higher energy capture rate and they use this captured energy more rapidly per gram. This appears to be approximately in relation to the surface to volume ratio (i.e., smaller organisms have more skin per gram of body mass). Another suggested reason is that the longer distance that body fluids must travel, the more energy it takes per gram.

An interesting consequence of this is that if you take all the organisms in the sea and estimate their biomass on a log scale there is as much biomass of large whales in the oceans as tuna fish as herring as krill as copepods as large phytoplankton as…. all the way down to tiny nannoplankton (Sheldon et al. 1972). Sheldon explains this as a consequence of each size, starting with the smallest, having higher production rates (that is a shorter time to replace its own biomass) than the next larger, and consequently can sustain the next larger organisms group while maintaining its own biomass. This, combined with the ecological efficiency to transfer food from one trophic (food, or here size) level to another, means that each level turns over fast enough to compensate for the loss of food from one level to the next.

In conclusion, all life is subject to the basic laws of energy, and all of life is about investing the energy it has into procedures to get more energy. The currency of energy in biology and elsewhere is electron flow operating in complex redox reactions. There are physical limits to what most life can do, most fundamentally the finite input of sunshine and the limitations imposed by surface to volume relations of different-sized organisms.

References

Brown, J.H., J.F. Gillooly, A.P. Allen, V.M. Savage, and G.B. West 2004. Toward a metabolic theory of ecology. *Ecology* 85: 1771–1789.
Dobzhansky, T. 1973. Nothing in biology makes sense except in the light of evolution. *The American Biology Teacher* 35(3): 125–129.
Falkowski, P. 2006. Evolution: tracing oxygen's imprint on earth's metabolic evolution. *Science* 322: 1724–1725.
Hatton, I.A., K.S McCann, J.M. Fryxell, T.J. Davies, M. Smerlak, A.R. Sinclair, and M. Loreau 2015. The predator-prey power law: Biomass scaling across terrestrial and aquatic biomes. *Science* 349(6252):p.aac6284.
Kleiber, M. 1932. Body size and metabolism. *Hilgardia* 6: 315–353.
Lehninger, A.L., D.L. Nelson, and M.L. Cox 2008. Principles of BioEnergetics Chapter 13 in Foundations of Biochemistry: http://www.bioinfo.org.cn/book/biochemistry/chapt13/bio5.htm.
Sheldon, R.W., A. Prakash, and W.H. Sutcliffe Jr 1972. The size distribution of particles in the Ocean. *Limnology and Oceanography* 27(3): 327–340.
Sibly, R.M., J.H. Brown, and A. Kodric-Brown 2012. Metabolic ecology: a scaling approach. Wiley-Blackwell, London, UK.
West, G.B., J.H. Brown, and B.J. Enquist 1997. A general model for the origin of allometric scaling laws in biology. *Science* 276: 122–126.

Chapter 6
Energy Return on Investment as Master Driver of Evolution

I repeat: "Nothing in biology makes sense except in the light of evolution." Theodosius Dobzhansky (1973). This chapter is about energy and evolution. We start with a brief review of Darwinian evolution, then build on that by first considering another important question in ecology relating to the distribution and abundance of species. I do this through a summary of some of the most important history on this issue.

6.1 Darwinian Evolution

Probably the greatest and most important conceptual contribution to our understanding of the world and our place in it comes from the thinking of Charles Darwin, who was an amazing observer and interpreter of nature. Probably most readers are familiar with the basics of his life which included a three year voyage around the world where he observed, collected, and thought about many amazing new things as the naturalist on the British Admiralty ship "Beagle" as it charted coastal waters in South America and elsewhere. Darwin used this time extremely well, traveling inland by horseback, living with Argentine Gauchos ("the happiest times of my life") and collecting plants, animals, and fossils.

Most important was the time he spent in the Galapagos Islands, a series of relatively small islands off the coast of Ecuador. There he observed and collected a series of finch (bird) species, similar to each other but different enough to be separate species. They were also similar to, but different from, mainland finches. In time he reasoned that one or a few pairs were blown from the mainland to the Islands, reproduced and modestly thrived. From time to time some of the birds were blown or traveled to other Islands of the group where, again over time, they became differentiated in accordance to the food and other possibilities of their new home. Some became eaters of large seeds, some of small seeds, some insects, one developed wood pecker habits using a cactus spine to pry insects out of trees and so on. Darwin reasoned that each species had developed from the ancestral form, and coined the phrase "descent with modification." While the idea of evolution, that is the concept that species could change (certainly evident to animal breeders who Darwin interacted with at length) was certainly "around," the incredibly important concept that Darwin added was "evolution with natural selection." He got this idea while reading Thomas Malthus' "Essay on Population" for pleasure one evening, where it became obvious to him that each species was generating far more offspring than would be needed to replace their parents. As a consequence most of these offspring had to die before reproducing or the earth's surface would have been buried long ago under exponentially increasing biomass of every species. Darwin's great insight was that on average it is only the most fit that would survive to reproduce. Darwin, a fairly shy and retiring person, did not wish to confront the enmity that his ideas brought forth in a world dominated by the perspective of the Church and the Bible, which held that species were unchanging and all had been created in one busy week by an all-powerful God. Fortunately his ideas were picked up by Thomas Huxley (sometimes called "Darwin's bulldog"), who took his ideas into the public where they gradually became dominant, at least amongst thinking people.

Darwin's Theory of Evolution as a consequence of natural selection is built on four principles:

- More individuals are produced than survive
- There is diversity (variability) in members of a population
- This variability is inherited
- There is differential survival (natural selection) and reproduction.

If these principles are true then natural selection and evolution MUST occur. I taught freshmen for many years, and it was easy to test these principles with them. It was obvious that the second and third conditions were true: even within one classroom there would be considerable diversity of student appearances, and students agreed that they tended to look more like their biological parents than others' parents. It was a little harder to test the first and fourth principle in modern humans, but I would ask the class how many would be alive without medical intervention and usually a third to half would raise their hands, so clearly if we were in "natural" conditions the first and fourth points would also be true. So it was pretty easy to convince a group of freshmen that natural selection was operating on themselves, and would be doing much more so without medical intervention. We then would discuss situations with hospital pathogens and agricultural weeds where practical people must deal with evolving pathogens and pests every day. If you have friends who do not believe in evolution have them talk to a hospital administrator or nurse struggling to keep ahead of evolving pathogens!

Evolution by natural selection is extremely important for understanding the world in that it gives a paradigm through which modern biology is understood and interpreted, as it operates as a organizing theme for all disciplines related to the biologic sciences (such as plant breeding and medicine) and it is crucial for understanding the environment and important issues related to the environment. But it cannot answer more existential questions such as the meaning of life beyond survival and reproduction, and maybe not even that in our overcrowded world. Early evolutionists such as Spencer believed that evolution was progressive, that led to "higher" forms terminating in humans. Modern biology eschews such goal orientation. In the words of ecologist Larry Slobodkin "Evolution is an existential game the object of which is to keep playing."

As time and science have marched on there is more and more evidence about the accuracy of Darwin's observations and interpretations. An amazing thing is as we understand more and more about biology at the molecular and biochemical level all we understand continues to be consistent with Darwinian principles. Darwin thought we could never see evolution, as it was too slow. But the work of Peter and Rosemary Grant showed that we could see the results of natural selection over just decades, even years. They went back to the Galapagos Islands many times, finding, as did Darwin, that they were the perfect laboratory for studying evolution by natural selection. The islands, located approximately on the equator, are very much subject to wet and dry periods due to the effects of the cycling climatic conditions known as El Niño and La Niña. The Grants measured the length and depth of the bills of one species of bird, a Darwin's finch, very carefully. During relatively dry periods plants that made small seeds were more abundant, and the size distribution of bills tended toward smaller bills. But during periods of relatively high rainfall plant species with large seeds tended to produce many seeds, and birds with larger bills were favored. Thus they found that over roughly 7 year cycles the bills of birds of this species would change from narrower and smaller, best for picking up the small seeds, and then in wet periods they would get larger and more robust, which were better for cracking the larger seeds! This was due to the differential survival and reproduction of birds that happened to be born with the finer or more robust bills, even over a relatively small number of generations. They observed and very carefully measured natural selection in action in Darwin's own bailiwick!

Darwin got the big picture absolutely right, and an astonishing fact that cements the importance of his concept is that as we understand biology deeper and deeper, especially with advances in biochemistry and molecular biology, what we find remains absolutely consist with basic Darwinian principles and our appreciation for the basic concept of natural selection as laid out by Charles Darwin becomes more profound.

Nevertheless there were certainly many important aspects of evolution that Darwin did not, indeed could not, understand. What are those aspects? The first is the work of Gregor Mendel that showed that one could not necessarily tell genotype (what genes are in an organism) from phenotype (what the organism looks like). This had puzzled Darwin considerably, as offspring did not necessarily inherit the parent's characteristics. Fundamentally Darwin did not (and could not) understand the role of dominant versus recessive genes (i.e., two recessive genes may combine to give offspring that look very different from parents). Nor did he (or could he) understand the role of population genetics, as described below. Finally he did not, and given the times could not, understand or appreciate the role of energy, which will be the topic of most of the rest of this chapter.

When I first took courses in genetics and evolution in the 1960s, the process of evolution by natural selection was explained to me approximately as above except that genes within a species were considered more or less fixed (with some variation, of course) and most variability was generated by "mutations," or chance changes in the genetic code, for example by a cosmic ray hitting a gene. In the intervening years one of the many things we have learned is that natural selection and evolution are not so much generated from such de novo gene changes. Rather most species populations (i.e., groups of individuals of the same species that are actually or potentially interbreeding) already have a great deal of genetic diversity and natural selection and evolution occur more by gene frequency shifts than by mutation. This is obvious in, for example, the work of the Grants mentioned above.

6.2 Fitness

Darwin, inspired by Malthus, noted that most organisms generated several to millions of times more offspring than were necessary to replace their selves, so that, given that most populations in nature were not increasing, the vast majority of offspring must die before reproduction. According to Darwin this is the basis for natural selection, since those organisms best adapted to their environment are more likely to survive and reproduce. But which ones survive to reproduce, and which ones die? Which ones reproduce the most, and hence provide a large genetic pool for natural selection to work on?

In 1864, British sociologist Herbert Spencer coined the phrase "survival of the fittest" (or "survival of the best fitted") for what Charles Darwin had called natural selection, and this phrase caught on. We can state this more precisely as "survival of the form (phenotypic or genotypic) that will leave the most copies of itself in subsequent generations." This concept should be viewed from a population perspective, as a probability. Thus the phrase "expected number of offspring" means the probable number in a population, not the precise number produced by one individual. Biology in the last century has examined the question of which organisms are favored mostly by focusing on the concept of fitness; that is on the ability of organisms to propel their genes into the future through continuation and expansion of populations of species. But what determines a specie's ability to survive and reproduce? The usual answer is fitness. But then how is fitness defined? The answer is "the properties of those individuals that are best at surviving and reproducing." The circularity of this argument has not gone unnoticed. I provide a better answer later, but first it is important to examine in more detail how evolution works in actual populations. I do this in the context of one of the most interesting and dominant themes of ecology during the past century: "What is it that determines the distribution and abundance of species?"

6.3 What Determines the Distribution and Abundance of Species?

The world and its biota are astonishing in their diversity and patterns of distribution. Each region has its own particularities of climate, geology, soil, and so on, and within this exists a suite of plants and animals that seem so perfectly adapted to the conditions of their particular region or regions. This applies even to the mind numbing total diversity of biotic life: each species seems to be perfectly designed to live where in fact it is found. Biologists answer that it is the interplay of environment and organism through natural selection that generates such well-designed species in each environment. But most species have a very restricted range. A good question is why are not more species much more widely distributed (like humans or many bacteria). We will address that problem by examining some important research in ecology that has asked that question, or more generally "what is responsible for species' distribution and abundance."

The earliest scientific ecologists were most interested in cataloging the species found in different parts of the Earth. Great early naturalists such as Alexander Humbolt, Alfred Wallace, and Charles Darwin made great expeditions all around the world, sending an amazing number and diversity of plants, animals, and fossils back to museums in their home countries. Initially they had little information to help them understand why different organisms were found where they were, and not elsewhere, although they were well aware that the distributions were far more than random. Humbolt in particular thought about the reasons for spatial distributions and even generated a map of the distribution of plant types along the geographical and elevational patterns of the Andes Mountains. But the reason for this distribution remained elusive for many years despite Humbolt's efforts to unite the physical and botanical sciences. He died in 1850, nearly a decade before Darwin's Origin of Species was published. Obviously much of the information that would have informed his efforts had not been made available although he had corresponded with Darwin.

Darwin himself wrote (not especially insightfully by later standards) about "Geographical distribution" in Chap. 11 of the Origin of Species. His main concern was how areas of similar climate ("physical conditions") but separated by geography or physical barriers often did not have the same species. His main concern was that appropriate species had not been able to migrate to the appropriate conditions in other locations.

We now understand much better how there is an interplay between dispersal mechanisms and the physical conditions that determine where different species are found. In the first half of the twentieth century, ecologists interested in the distribution of species tended to focus on how different groups of species of plants (and sometimes animals) tended to be found together as "associations" or "super organisms" each in their own climatic conditions—for example "beech–maple" or "oak–hickory" forests. The "plant community" concept, that species somehow facilitated each other's existence, was put forth by Frederick Clements who dominated thinking in plant ecology for many years. His perspective that various species would reinforce each other and even create conditions facilitating their own well-being was consistent with a certain "holistic" view of nature that was then popular. Standing against that view was Henry Gleason, who believed that each species was distributed along gradients (changes) in environmental conditions independent of other species. This perspective tended to have less influence, at least initially.

This distribution pattern is seen perhaps most clearly with variations in temperature. Brook trout are well adapted to cold water, and will not be found in warm water. Likewise largemouth bass are found in warmer water but not cold. But brook trout are often found together with black nose dace but not largemouth bass. Is there some connection? Tree species also are very sensitive to temperature and each species has a distribution founded by a maximum and minimum temperature that it can tolerate. Likewise different tree species are found differentially according to soil water availability. But are these species, when not influenced by humans, found independently, as Gleason suggests, and only coincidentally with particular other species who share part of the same range, or are they found in associations as suggested by Clements? In a classic study of tree vegetation the great plant biologist Robert Whittaker tested the Clementsian view versus the Gleasonian view by hiking up and down the Smokey Mountains in Tennessee and carefully mapping the distribution of various tree species he encountered there. He found, in accordance to Gleason's hypothesis, that each species tended to have an independent and relatively narrow range of elevation (and hence temperature) where it would be abundant. Oaks and pines were found in drier sites at lower elevation, where it was warmer, but each species were found together only incidentally. At higher elevations spruces and firs were found. Likewise some species, such as red maples, were abundant in wet environments, but did not do as well in drier regions where chestnut and oaks were more abundant (Whittaker 1956). In each case species were found in what might be called associations only where their independent ranges partially overlapped. A special case in certain environments (usually microenvironments) is that some organisms do in fact co-occur, in special symbiotic circumstances, but this is rare compared to species each adjusting independently to their environment.

A very interesting issue is "why the relatively narrow range in each specie's distribution"? Why are not firs found from the bottom to the top of the Smokey Mountains? Why are oaks found only on drier sites, instead of also where soil water is more abundant? The answer seems to be that there are *opportunity costs* or *tradeoffs* to living in each particular region. Enzymes, for example, are not always "on" or functioning, but operate (and should operate) only in specific conditions, such as appropriate pH, temperatures, existence of other proteins and so on. An obvious example is our own gut enzymes. If digestive enzymes were always operating they would be digesting our own gut lining. Instead we can digest food while protecting the gut lining. There is an abundance of information about how there are tradeoffs in the use of enzymes, for example, at one temperature versus another (see the many examples given in Hall et al. 1992, including Powers 1986 and the summaries in Hochachka and Somero 1984). Zhou et al. (2014) test the response to drought of different tree species in Australia and Spain that are characteristic of mesic (moist) and xeric (dry) environments and find that there are tradeoffs: being optimally adapted to a wetter environment has costs relative to being adapted to drier environments and the converse. Thus in general there has been natural selection for organisms to optimize their use of a given set of environmental conditions that make them suboptimally adapted to other environments. In other words species appear to be groups of morphologies, physiologies, and behaviors that are optimally adjusted for one set of environmental conditions, and of necessity suboptimally adapted to other environments. Different species can be found as strays or small populations along a large range of environmental conditions, but each species will be abundant only in the middle of its range where the environmental conditions are optimal and reproduction is large (Hall et al. 1992).

This perspective can account for much of what we observe in nature, but still leaves unanswered the mechanism by which the different morphologies, physiologies and behaviors are translated into species' relative success or failure at different environmental conditions. Nor have we answered yet how these differential responses are translated into evolutionary advantage. For this we turn to energy, and many believe that the total energy balance of an organism is the key to understanding fitness. Explicitly, Hall et al. (1992) found that varying energy costs and energy gains in response to environmental gradients explained where each species could, or could not, make a sufficient energy profit to survive and reproduce. This provides a non circular definition of fitness: fitness is greatest where and when the difference between energy costs and energy gains is greatest, providing a large energy surplus that can be, and usually is, translated into survival and reproduction.

6.4 Energy as the Master Resource for Evolution

The central role of useful energy in ecology and evolution was appreciated by some early scientists. Ludwig Boltzmann (1905) wrote, "The 'struggle for existence' of living beings is not for the fundamental constitutes (elements) of food, which are everywhere present in earth, air, and water, nor even for energy, as such, which is contained, in the form of heat, in abundance in all bodies, but for the possession of the free energy obtained, chiefly by means of the green plant, from the transfer of radiate energy from the hot sun to the cold earth." Alfred Lotka also agreed (1922): "The life contest, then is primarily a competition for available energy." Now it is possible to start placing major features of ecology and evolution into a quantitative predictive theory based explicitly on energy and metabolism.

While Lotka is (curiously) much more well known for the cycling "Lotka Volterra curves" (think Hare and Lynx, but see Hall 1988 for debunking), he personally thought his energy insights much more powerful. Leigh Van Valen (e.g.1976) was a very thoughtful evolutionary ecologist who thought deeply about energy and natural selection. He is known in particular for the "Red Queen hypothesis" (from Lewis Carroll's Through the Looking-Glass) where the Red Queen must run faster and faster to stay in the same place. He thought that this also applied to natural selection where organisms are constantly evolving to become better adapted to their environment—but other species are doing that at the same time!

Defining fitness in energetic terms avoids the circularity previously noted and explicitly incorporates the central role of energy in two universal characteristics of living things: the potential for exponential population growth and evolution by natural selection. Each of these processes combines ecology and evolution, and together they form the Malthusian–Darwinian dynamic (MDD; Burger et al. 2012). Plants and animals are subjected to fierce selective pressure to do the "right thing." It is increasingly clear that those surviving and reproducing organisms, usually considered "the most fit," are those that energetically have the right morphologies, physiologies and behaviors to ensure that whatever major activity that they invest in gains more energy than it costs, and beyond that gets a larger energy net return than either alternative activities or their competitors. This is the "iron law of energy" introduced in the last chapter. If they are to prosper through time organisms must extract energy from the environment, process it within their bodies, and allocate it to maintenance (survival) and production (growth and reproduction, including surplus reproduction upon which natural selection can operate).

Organisms must capture and use energy to exist. The fundamental reason is that life requires a very specific arrangement of elements and molecules according to the DNA "plan" of the organism. In the absence of energy applied to the maintenance of the biotic structure of the molecules according to the plan, the molecules will tend toward randomness, a process sometimes known as "entropy." Thus at the minimum organisms must capture energy or they will cease to exist—as is obvious to us from the putrefying appearance and smell of a dead organism. Thus organisms are a member of a general class of organization called *"dissipating structures"*—they are entities which are characterized by exploiting and then using energy, dissipating the free energy into heat. Other examples of dissipating structures include stars, hurricanes, and rivers. Each of these maintains structure by extracting energy from the environment and using it to build, maintain, and grow structure. And, to answer our question about distribution, species are genetic packages of morphologies, physiologies, and behaviors that are optimally suited to do this at their own locations along diverse environmental gradients.

6.5 Energy Return on Investment as Master Driver

Equally important, organisms must gain more energy than they use in obtaining it. We can say more formally that organisms must have a positive EROI, or energy return on investment. EROI is formally defined as a ratio:

$$\text{EROI} = \frac{\text{Energy returned from an energy gathering activity}}{\text{Energy used to get that energy}}$$

This basic equation integrates the two main themes of this book, investments and energy, and provides a means to understand very many things that are going on in the biological and, as we shall see, human economic worlds. Next we give a number of examples of how EROI operates with familiar examples from the biological world, and then show how it can be used to give a less circular definition for evolution. If you examine natural organisms in their own environment they are usually feeding, positioning themselves to be able to feed, or resting in between feeding bouts. In other words the activities of most organisms (except some times in breeding season) consist of investing energy to obtain more energy (usually as food) or to avoid additional energy losses. Many studies of organisms in natural environments have examined the relation between the energy invested in getting food and the energy gained by obtaining it. It is obvious that a cheetah, for example, has to catch more energy in its prey than it takes to stalk it and run it down (and only about 10 % of their chases are successful!), and considerably more to make it through lean times and also to reproduce and feed the kittens. Animal migrations are a large investment that allow organisms to exploit spatially and temporally varying environments (e.g. Hall 1972; Thomas et al. 2001). Plants too must make an energy profit to supply net resources for growth and reproduction. This can be seen easily in most clearings or roads in evergreen forests, where living boughs on a tree that are in the clearing side are usually lower down than they are in the more densely forested, and hence shaded, side of the tree (Figs. 6.1 and 6.2). If you look at the trunk on the shaded side you will see dead branches, or stumps, or small circles where there were once boughs opposite the living ones on the open side, but which are no longer there. Obviously, if the bough does not carry its weight energetically, that is if its photosynthesis is not greater than the respiratory maintenance metabolism of supporting that bough, the bough will die (or perhaps be sloughed off by the rest of the tree).

Trout in streams show this operating in real ecosystems (e.g., Li and Brockson 1977; Smith and Li 1983). Trout in streams are mostly drift feeders, holding one position and feeding on insects that drift by with the current as they undertake their life cycles or are blown into the stream. A trout in very fast water will have lots of food drifting past, but its energy costs will be very high, reducing the net energy obtained. A trout in slow water can be very efficient, turning a high proportion of the food it gets into growth of tissue or gametes, because its swimming costs are low, but the slower water brings with it less food. The net energy supplied to the fish is the difference between the energy invested in swimming and the energy gained from the number of insects drifting past. Smith and Li, in a series of beautiful field studies and experiments, showed that dominant trout will pick an intermediate, optimum, current speed, which results in faster growth and more offspring. Subdominant trout will be found in water moving a little faster or a little slower. In some experiments trout with no competitive power will be found drifting aimlessly in still water slowly starving to death. These relations are also obvious to an observant fly fisherman who knows that generally the largest trout are found neither at the fast water at the top of the pool where the riffle pours in, nor in the slack water in the middle of the pool, but at an intermediate rate of water flow.

6.5.1 A Special Section for Fly Fishers

I have spent considerable part of my life from pre-high school to now pursuing trout and salmon with a fly rod. As any fly fisherman or woman knows the fishing is

Fig. 6.1 Energy costs and gains are obvious to the casual walker along a path in evergreen forests where a shaded tree limb (on the *right-hand side* of the *left-hand tree*) that cannot pay for its metabolic cost is sloughed off

pretty slow much of the time, as one waits for the insects or the trout to become active (after all, brown trout, especially, will not crank up their metabolism, wasting energy, if there are not sufficient gains from an adequate supply of insects). This gives the fly fisher lots of time to contemplate—and, if one is such as I, predisposed, to think about energy. Thus I have thought a lot about aquatic insects, trout, and energy. Here are a few conclusions I have, presented here outside the world of rigorous journal articles.

First, in areas such as New England or the Catskill Mountains of New York there is a regular progression of insect species with the seasons: Quill Gordons and Hendricksons in Mid-April, Grey Foxes and March Browns in May, and Light Cahills and Cream Variants in June (and many others). The evocative names often reflect early Catskill fly tiers. Much care has gone into the tying of these patterns, and what fly tier can forget that a proper Catskill Hendrickson requires "urine-stained belly hair from a vixen (female fox)" as dubbing for a proper body? The early flies, the Quill Gordons and the Hendricksons, are dark and "hatch" or emerge from their underwater larval state in early afternoon—we used to say that you could set your watch at 1:45 when the first Hendricksons would appear. As the season progressed the flies would be later in the day and lighter colored, until the Light Cahills and Cream variants of June emerged at sunset. Obviously the earlier flies had to absorb as much sunlight as possible to get their wings warmed up enough to take off from the water and avoid the hungry trout. Why the flies in June are light colored is not as easily answered—perhaps their lighter silhouette is not as visible to trout in the late evening light.

6.5 Energy Return on Investment as Master Driver

Fig. 6.2 Another example of boughs that are not shaded by other trees being kept by the tree and limbs that are shaded by another tree (or structure) dying and being sloughed off. A net energy profit is a powerful and fearsome process in nature

But mayflies have some tough decisions too. As with most species the females make the choice of mates. They have spent their lives yoking up their eggs, a very energy-intensive process. They have not been selected to throw this investment away on bad male genes (an issue that confronts females of our species too). If you watch along the clean rivers and lake sides as the sun goes down you will see mayflies dancing beautifully in swarms, flying up and then floating down gracefully. If you put a butterfly net through the swarm it will be all males (large eyes, long forelegs). After about 45 min the females will appear, and they will fly at full tilt through the swarm of males. A number of males will notice and a mad pursuit ensues. Eventually the "best" male catches the female, and gets to fertilize her eggs. Why do female mayflies (and one might say females in general) subject the males to such an arduous mating ritual? I suspect that they have been selected to find which male "has the most gas in his gas tank," i.e., is the best forager, meaning the one who has maximized his energy return on investment.

Finally I have some messages to younger fly fishermen and women: I suspect getting out in beautiful nature and learning the intricacies of water flow, insect and trout life histories and so on are as fascinating to you as they were to me. But things have changed so much, mostly with a hundred times more fishermen converging on the banks of the same poor streams, poorly mannered "me generation" people not respecting other fishermen who may have been there first, and especially the tranformation of fishing from a quiet, cheap personal activity into another commoditized, industrialized, frequently privatized economic activity of expensive guides, lodges and drift boats. This includes their expensive, energy intensive backup services that frequently dominate the rivers and riverside roads, increasingly dotted with MacMansions and "keep out" signs. Fishing, like so many things, has become commoditized, and the younger fishers might never know what they have lost. Meanwhile I avoid the blue ribbon rivers I used to love, happy while my legs last to bushwhack into bear country to bring back a few modest trout for supper.

6.5.2 EROI and the Growth of Tits

Thomas et al. (2001) showed even more explicitly how powerfully net energy controlled fitness. They studied tits (chickadees) in France and Corsica using double-labeled isotopes that could measure both energy acquisition and use. More explicitly they studied the timing of migrations and the energetics and success of nesting and feeding their young with exquisite experimental procedures. They found that those birds that timed their migrations, nest building, and births of their young to coincide with the seasonal availability of large caterpillars, which in turn were dependent upon the timing of the vernalization of the oak leaves they fed upon, had a much greater surplus energy than their counterparts that missed the caterpillars. They fledged more, larger and hence more-likely-to-survive young while also greatly increasing their own probability to return the next year to breed again. Those of their offspring that inherited the proper "calendar" for migration and nesting were in turn far more likely to have successful mating and so on. Thomas et al. also showed how the natural evolutionary pattern was being disrupted by climate change, so that the tits tended to get to their nesting sites too late to capitalize upon the caterpillars, who were emerging earlier in response to earlier leaf out. The adults worked themselves into a frazzle trying to feed their young, and fewer young or adults survived to the next year. Presumably if and as climate warming continues natural selection will favor those tits which happened to have genes that told them to move North a bit earlier.

Pathogens too impose an energy loss on organisms even when they do not kill them. Moret and Schmid-Hempel (2000), in a particularly nice study, trained bumblebees to feed off of small glass spheres, which the bees mistook for pollen. When the bees were fed this diet they would die from lack of energy in about 5 days. When the investigators infected the bees with a bumblebee pathogen the bumblebees would survive if they had real food but would die in only 3 days when fed the glass spheres. This shows that when challenged with pathogens the bumblebees need to use their own energy reserves to fight them.

Of course life in all of its diversity also has a diversity of energy life styles that have been selected for—sloths are just as evolutionarily successful as cheetahs, while warm-blooded animals pay for their superior ability to forage in cold weather with a higher energy cost to maintain an elevated body temperature—the list is endless. While drift-feeding trout choose areas of intermediate current to maximize the energy surplus, suckers have "chosen" through natural selection (i.e., have been selected for) to maximize energy surplus by processing lower quality food on the

6.5 Energy Return on Investment as Master Driver

bottom and using less energy in swimming, and probably have an optimum power output for that set of environmental conditions.

Nevertheless each life style must be able to turn in an energy profit sufficient to survive, reproduce, and make it through tough times. There are few, if any, examples of extant species that barely make an energy profit—for each has to pay for not only their maintenance metabolism but also their "depreciation" and "research and development" (i.e., evolution), just as a business must, out of current income. Thus their energy profit must be sufficient to mate, raise their young, "pay" the predators and the pathogens and adjust to environmental change through sufficient surplus reproduction to allow evolution. Only those organisms with a sufficient net output are able to undertake this through evolutionary time, and indeed some 99+ % of all species that have ever lived on the planet are no longer with us—their "technology" was not adequate, or adequately flexible, to supply sufficient net energy to balance gains against losses as their environment changed. Given losses to predation and pathogens, nesting failures and the requirements of energy for many other things, the energy surplus needs to be quite substantial for the species to survive in time. Each investment in each contingency has an energy opportunity cost, as it diverts an organism's available physiological or behavioral energy and leaves less for other investments. Consequently most species are abundant in only a small part of the environmental conditions in which they might be found.

6.6 Energy Return on Investment as the Means of Obtaining Darwinian Fitness

Thus the most general and non-circular definition of fitness is in terms of energy return on investment (EROI). The differential fitness of individuals and populations is due to positive, and probably high, realized energy return from energy invested. The potential to realize a positive EROI is the basis of success within the Malthusian–Darwinian dynamic. Applied to heritable traits, EROI provides a measure of the strength of a particular genome for natural selection. A large EROI allows the organism to withstand the opportunity tradeoffs as the organism adjusts to the various energy-consuming contingencies that it faces even as it fine-tunes its genome to a specific location along a series of gradients.

While I am unaware of any official pronouncement of this idea as a law, it seems to me to be so self-obvious that we might as well call it a law—the "iron law of energy" or perhaps "the law of maximum EROI." Every plant and every animal must conform to this iron law of evolutionary energetics: if you are to survive, you must produce or capture more energy than you use to obtain it, if you are to reproduce you must have a large surplus beyond metabolic needs, and if your species are to prosper over evolutionary time you must have a very large surplus for the average individual to compensate for the large losses that occur to the majority of the population as less fit or unlucky organisms are weeded out. In other words, every surviving individual and species needs to do things that gain more energy than they cost, and those species that are successful in an evolutionary sense are those that generate a great deal of surplus energy that allows them to become abundant and to spread and to recover from emergencies, such as adverse weather events. While probably most biologists tacitly understand and even accept this law (if they have thought about the issue) it is not particularly emphasized in biological teaching.

The genetic diversity that exists in most species can be explained in part due to the fact that not all genomes of a species are equally fit along the set of environmental conditions (gradients) where that species is found. While each individual is sufficiently fit, that is makes a sufficient net energy gain, where it is found, that might be somewhat different from one part of its range to another. Generally it would not be so fit in the different conditions of a different location (e.g., warmer or colder, or wetter or drier, or in some way different), because its particular genome is not quite appropriate for that nonoptimal location. But probably another species, honed in on this new location, would be better suited, i.e., more fit, i.e., having a higher EROI there. This is an explanation for species diversity: different species are differentially adapted to each location in gradient space, at a cost of being less well adapted for another position along the gradients. The net result is a world of individual species, in accordance to Gleason's views, each being specialized for a different set of environmental conditions. The result is the marvelous diversity of the world we know.

References

Burger JR, Allen CD, Brown JH, Burnside WR, Davidson AD, Fristoe TS, Hamilton MJ, Mercado-Silva N, Nekola JC, Okie JG, Zuo W (2012) The macroecology of sustainability. PLoS Biology 10(6):e1001345.
Dobzhansky T (1973) Nothing in Biology Makes Sense except in the Light of Evolution The American Biology Teacher, Vol 35, No. 3 (March, 1973), pp 125–129.
Hall CAS (1972) Migration and metabolism in a temperate stream ecosystem. Ecology 53(4): 585–604.
Hall CAS (1988) An assessment of several of the historically most influential theoretical models used in ecology and of the data provided in their support. Ecological Modeling 43:5–31.
Hall CAS, Stanford JA, Hauer R (1992) The distribution and abundance of organisms as a consequence of energy balances along multiple environmental gradients. Oikos 65:377–390.
Hochachka PW, Somero GN (1984) Biochemical adaptation. Princeton University Press.
Li HW, Brockson RW (1977) Approaches to the analysis of energetic costs of intraspecific competition for space by rainbow trout (Salmo gairdneri). J Fish Biol 11:329–341.
Moret Y, Schmid-Hempel P (2000) Survival for immunity: the price of immune system activation for bumblebee workers. Science 290:1166–1168.
Powers DA (1986) The use of enzyme kinetics to predict differences in cellular metabolism, development rate and swimming performance between LDH-B genotypes of the fish Fundulus heteroclitus. Curr Topics Biol Med Res 10:147–170.
Smith JJ, Li HW (1983) Energetic factors influencing foraging tactics of juvenile steelhead trout, Salmo gairdneri. In: Predators and prey in fishes (ed) Noakes DLG et al. 173–180. Dr W. Junk Pub. The Hague.
Thomas DW, Blondel J, Perret P, Lambrechts MM, Speakman JR (2001) Energetic and fitness costs of mismatching resource supply and demand in seasonally breeding birds. Science 291:2598–2600.
Van Valen L (1976) Energy and evolution. Evolutionary Theory 1:179–229.
Zhou Shuangxi (2014) Belinda Medlyn, Santiago Sabaté, Dominik Sperlich and I. Colin Prentice. 2014. Short-term water stress impacts on stomatal, mesophyll and biochemical limitations to photosynthesis differ consistently among tree species from contrasting climates. Tree Physiology 34(10):1035–1046.

Chapter 7
Maximum Power and Biology

Neither the first nor the second laws of thermodynamics address the *rate* at which energy transformations or processes occur. But the rate of energy use is critically important, as is obvious in a foot race. In a competitive world, it is important to not only "out energy" one's competition but to do it fairly rapidly, that is before the resource is captured by another individual or species. Power measures the rate at which energy is used. The concept of maximum power incorporates time into measures of energy transformations. It provides information about the rate at which one kind of energy is transformed into another as well as the efficiency of that transformation. As such it is critical to understanding the role of energy in evolution.

For many years I struggled with, or ignored, the relation between EROI and maximum power, both large ideas about energy and biological or social evolution. They were both fascinating to me, just too different. Recently I think I have come up with a resolution: I now think of maximum power as a subset of EROI, it is a means of optimizing EROI by paying attention to time. Both maximum power and its relation to EROI will be developed more fully below.

7.1 History

In his book "On the Origin of Species" Darwin proposed that organisms were involved in a life struggle over scarce resources, and that the process of natural selection caused new varieties and new species to emerge. Ludwig Boltzmann proposed that the primary object of contention in this life struggle was *available energy* and Lotka and others (1922a, 1922b) proposed that natural selection was, at its root, a struggle among organisms for available energy; that organisms that survive and prosper are those that capture and use energy with greater efficacy and more optimal efficiency than their competitors. In other words energy is a general resource that can be diverted to whatever contingencies an organism faces, and the efficient accumulation of energy allows maximum reproductive output which is, after all, what natural selection is based on. "By this theory organisms (or even systems of cycles) that drain less energy lose out in competition to those that drain more energy from their environment." Lotka's statements sought to explain the Darwinian notion of evolution as being based, in part, on a very generalized and strong physical (energy) principle.

Curiously Lotka is much more well known for the population cycling "Lotka Volterra" curves (think of the well-known Canadian hare and lynx cycles over time based on pelt returns to Hudson Bay's company—but see Hall 1988 for a debunking of that example; hares were from Eastern Canada and Lynx from Western Canada). Lotka personally thought his energy insight much more important and powerful than the population approach for which he is most known.

In his paper "Contribution to the Energetics of Evolution" (1922a) Lotka moved the discussion even further, from the energetics of a single organism or species to the energetics of entire energy pathways through ecosystems. After examination of several situations he said "In every instance considered, natural selection will so operate as to increase the total mass of the organic system, to increase the rate of circulation of matter through the system, and to increase the total energy flux through the system, so long as there is presented an unutilized residue of matter and available energy. This may be expressed by saying that natural selection tends to make the energy flux through the system a maximum, so far as compatible with the constraints to which the system is subject" (Lotka 1922a, p. 148).

In a second paper, published simultaneously, Lotka (1922b, p. 151) further elaborated: "The principle of natural selection reveals itself as capable of yielding information which the first and second laws of thermodynamics cannot furnish. The two fundamental laws of thermodynamics are, of course, insufficient to determine the course of events in a physical system. They tell us that certain things cannot happen, but they do not tell us what does happen."

The statement of maximum power as a formal principle was first proposed by Lotka (1922a, b), who thought that time, and hence power, was important: the theory of natural selection operated as a maximum power organizer of systems, and also that it applied to humans: "The question was raised whether, in this, man has been unconsciously fulfilling a law of nature, according to which some physical quantity in the system tends toward a maximum. This is now made to appear probable; and it is found that the physical quantity in question is of the dimensions of power, or energy per unit time…".

In summary both Boltzmann and Lotka proposed that natural selection was, at its root, a struggle among organisms for available energy; organisms that survive and prosper are those that capture and use energy at a rate and efficiency more effective than that of their competitors, in other words energy is a general resource that can be diverted to whatever contingencies an organism faces, and the maximum accumulation of energy allows maximum reproductive output which is, after all, what natural selection is based on.

7.2 Maximum Power for One Process

Lotka's concept was subsequently developed and greatly expanded by the systems ecologist Howard T. Odum, who initially started with wondering why the efficiency of energy transfer through food webs in ecosystems, which he had been studying intensely, was so low. This led him to a collaboration with the physicist Richard C. Pinkerton, initially on physical systems. They wrote a remarkable and fascinating paper "Time's speed regulator" (Odum and Pinkerton 1955). They started with the extremely simple "Atwood's machine," a pulley with a rope over it attached to two baskets (Fig. 7.1). The objective of the machine is to use a heavier weight on one side to move other weights up to the top of the system. One can imagine using elevated rocks to move coal (or gold) from an underground mine to the surface. One can move the coal most rapidly by having a large weight differential in the two baskets—the coal will zip to the top—but not much will be delivered and most of the input energy will end up as heat when the rapidly moving downward basket hits the ground. Alternatively, if the weights are nearly the same, much coal will be delivered—but very slowly. The maximum useful work (maximum useful power) is done when the input energy, the force (weight) of the elevated rocks, is about twice that of the load—i.e., the delivered load—the coal, and about half the input energy is lost as heat (Fig. 7.2). The authors continue with many examples portrayed in clever small diagrams for biochemical, electrical, and economic examples—some of the cleverest diagrams I have ever seen and well worth the reading of the original paper (Figs. 7.3, 7.4, 7.5, 7.6, 7.7, 7.8, 7.9, 7.10, 7.11 and 7.12). Odum and Pinkerton believed that the concept had very large application in any competitive system, and over the years Odum developed or reported on its application to many very different physical, biological, and economic systems. The concept appears to be very general, for example large coal-burning electrical plants operate at about 40 % efficiency when they are capable of operating at about 80 % efficiency, but to do so would not be economic because the plants would have to work infinitesimally slowly (Curzon and Alborn 1975). Other examples, such as trout feeding in water of different velocities, were given in the previous chapter. The underlying maximum power philosophy, here and elsewhere, aims to unify the theories and associated laws of physical, electronic, and thermodynamic systems with biological and economic systems.

Thus Odum argued from the ideas first proposed by Lotka that it is not just the net energy obtained but the power, that is the useful net energy per unit of time, that is critical in an evolutionary context. He stated the maximum power principle explicitly as: "During self-organization, system designs develop and prevail that maximize power intake, energy transformation, and those uses that reinforce

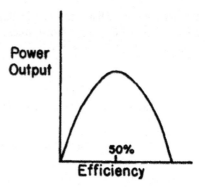

FIG. 1. Atwood's Machine. The power input due to the falling weight M_1 drives the elevation of weight M_2 which stores potential energy.

FIG. 2. Power output is given as a function of efficiency for systems where there is no leakage (l) and the efficiency (E) is thus equal to the force ratio (R).

Figs. 7.1 and 7.2 Atwood's Machine. The power input due to the falling weight M_1 drives the elevation of weight M_2 which stores potential energy. Power Output as a function of efficiency, which is inverse to the rate of a process (original figure). The *left hand of the equation* represents a high rate, low efficiency situation such as when $M_1 \gg M_2$. In this case most of the input energy goes into heat when the fast moving M_1 hits the basket at high velocity. The middle of the graph, where M_1 is twice M_2 is where power (useful work/time) is maximum. On the *right-hand side* is highest efficiency but the lowest rate, occurring when M_1 is barely greater than M_2

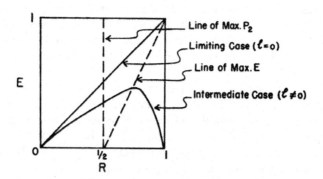

Fig. 7.3 Relation of efficiency (E) and force ratio (R). A typical curve is drawn for the case where leakage (l) is not zero. E values for maximum power output are located on the *vertical dashed line* where R is ½. Maximum efficiency and associated R values are found on the *slanting dashed line*

production and efficiency" (Odum 1995, p. 311) and "over time, through the process of trial and error, complex patterns of structure and processes have evolved… the successful ones surviving because they use the materials and energies well in

7.2 Maximum Power for One Process

FIG. 4. Water wheel driving a grindstone.

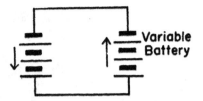

FIG. 5. One battery charging another battery.

Figs. 7.4 and 7.5 Examples of systems for which maximum power can be related to load, from Odum and Pinkerton

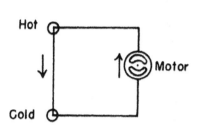

FIG. 6. Thermocouple running an electric motor.

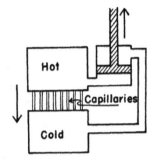

FIG. 7. Thermal diffusion engine; heat source driving a piston.

Figs. 7.6 and 7.7 Examples of systems for which maximum power can be related to load, from Odum and Pinkerton

their own maintenance and compete well with other patterns that chance interposes" (Odum 1983).

Maximum power was first stated explicitly in terms of the tradeoff between rate and efficiency of a single process; that is, the more rapidly a process occurs the lower its efficiency, and vice versa. Under a given set of environmental conditions, it is not selectively advantageous for an organism in a competitive world (as is the case for most of nature) to be extremely efficient at the expense of the rate of exploitation, nor to be extremely rapid at the expense of efficiency. An organism that operates efficiently, but slowly, may find that the resource it is seeking has been preempted already by a more rapid organism. Many examples can be found in normal everyday life. For example, if you want to accelerate on a bicycle (or automobile) you can do the most work (acceleration) in the middle of the appropriate gear range, not at the more efficient lower range or the more rapid upper portion (this can be seen in acceleration graphs in sports car magazines) (Fig. 7.13). Likewise when using a chain saw one does the most work by pressing down an intermediate amount on the log with the saw: too little pressure and the saw runs rapidly but most of the energy in the gasoline is transferred to heat, not work

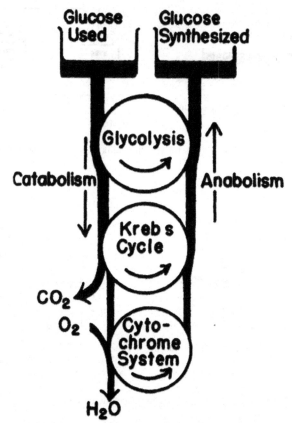

FIG. 8. Pseudo-organism; glucose catabolic enzyme system driving a glucose anabolic enzyme system.

Fig. 7.8 Example from Odum and Pinkerton

(Fig. 7.14). Push down too hard and the saw slows down (and eventually stalls). The most useful work is done at intermediate pressure of the saw onto the log.

In a similar, very general conceptual assessment it seems obvious that nature is often structured by maximum power: for example if one examines the leaf area index of broad leafed forests, which is the square meter of leaves per square meter of ground. One might think that there would be natural selection for a very high rate of photosynthesis, in this case a high leaf area index, say 12 m^2 of leaves per square meter of ground, since each additional increment of leaves would add more energy to the tree, presumably increasing its potential energy gain and hence fitness. But in fact leaves and branches are expensive to maintain, and each additional layer of leaves adds more photosynthesis but also more respirational energy cost. Plotting the photosynthesis and the respiration versus leaf area index gives a maximum difference between energy gains and energy costs at a leaf area index of about 6,

7.2 Maximum Power for One Process

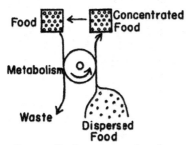

FIG. 9. Food concentration by an organism for its maintenance. Metabolism of concentrated units of captured food drives the process of capturing food.

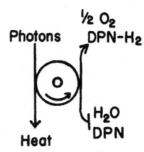

FIG. 10. Model of Photo-synthesis; Absorption of photons drives the reduction of coenzymes.

Figs. 7.9 and 7.10 Further examples from Odum and Pinkerton

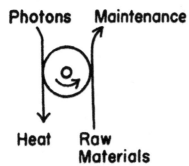

FIG. 11. Climax community. The absorption of photons drives the processes of growth necessary to maintain and repair the community.

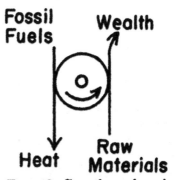

FIG. 12. Growth and maintenance of a civilization. The use of fossil fuels in power plants, etc., drives the production of wealth both for maintenance and growth.

Figs. 7.11 and 7.12 Further examples from Odum and Pinkerton

which is in fact what we find throughout the world's forests where there is sufficient water. In what appears to be an analogy from the physical world Leopold and Langbein (1966) have shown that streams, in developing erosion profiles, meander systems and tributary networks disperse their potential energies more slowly than if their channels were more direct—in other words they develop structure to capture and hold on to available energy by physical inanimate processes. Odum (1994, 1995) has argued that hurricanes and even stars operate to extract energy from their environment, build structure that maintain themselves while extracting more energy

Fig. 7.13 Maximum power is obvious to a bicycle rider who wants to stay in the middle of a given gear rang to generate maximum power for acceleration or climbing a hill

Fig. 7.14 Maximum power can be readily understood if you are familiar with using a chain saw. If you rest the saw blade lightly on the log the motor will run very fast but little work will be done (*left side* of Fig. 7.2). The energy of the gasoline will be converted almost entirely to heat. If you press down harder more chips will fly, but if you push down too hard the saw will slow down and eventually stall (*right side* of Fig. 7.2). The maximum amount of useful work (cutting the log) is done when you put intermediate downward pressure on the saw blade

7.3 Maximum Power for Systems

We have tested the maximum power hypothesis explicitly in forests along an elevational gradient in Puerto Rico (Harris et al. 2013). We had hypothesized that ecosystems developed the most useful power (net photosynthesis) at intermediate elevations. Gross photosynthesis was maximum at sea level, but there respiration also would be maximum too due to high temperatures. At high elevations photosynthesis is relatively low because of lower sunlight due to cloudiness, but so is respiration because of lower temperature. We developed procedures for measuring photosynthesis and respiration of a column of forest (including all species that were there) using a LI-COR CO_2 analysis machine and a giant slingshot and rock climbing technology to reach upper portions of the forests (Fig. 7.15). We undertook this at 11 different elevations from the base of the mountain to the summit. We found a clear tradeoff between rate and efficiency and a maximum net photosynthesis (useful power) at an intermediate elevation (Figs. 7.16 and 7.17). We wonder whether the environmental conditions at 800 m elevation in the Luquillo Forest represent something of ideal conditions for the balance of photosynthesis and respiration for trees more generally as forest net production also may show a maximum at intermediate latitudes (Huston and Wolverton 2009; but see Gillman et al. 2015). The lowland tropics are in many ways favorable to plant growth but high temperatures and long nights there impose a huge energy tax on both natural and cultivated ecosystems (Hall 2000).

Later Odum said that maximum power was not simply about the tradeoff of rate and efficiency of a particular operation but that it encompassed any procedure that led to a more effective capture of energy by an organism or trophic level. Specifically he said (From Hall 1995):

> There is selection for autocatalytic (self-reinforcing) processes that reinforce production. This would include the investment of energy within hurricanes that build structure (the vertical up moving air column) or ecosystems that generate more plant biomass that in turn captures more energy from their environment.
>
> Energy transformations are hierarchical such that maximum power at one level may reinforce patterns at larger or smaller scales.
>
> Energy transformations converge spatially and interact multiplicatively
>
> Production pathways generate their own storage which reinforce exploitation of more energy from the environment. (a hurricane is a good example)
>
> All systems pulse and these processes interact with the pulses. Odum believed that most systems would have cycles of production and destruction, such as periodic fires, floods or other natural "catastrophes" that would periodically release nutrients and allow for regeneration, with a consequential increase of overall productivity.

Fig. 7.15 How to access the upper leaves in a large tropical tree

In other words Odum thought that, in general, systems (including ecosystems) extracted energy from the environment and invested this energy in generating structure (such as the vertical column of a hurricane or the biomass of an ecosystem) that would extract more energy from the environment up to the limits imposed by the cost and efficiency of that extraction. He saw, for example, industrial society as a perfect example, developing structure to extract and utilize more fossil fuels up to the limit of the resource. He felt that societies or groups that chose not to do this could be more efficient but would be overtaken by systems that continued to focus on increasing their rate of exploitation of fossil fuels, at least while fuels were abundant.

7.3 Maximum Power for Systems

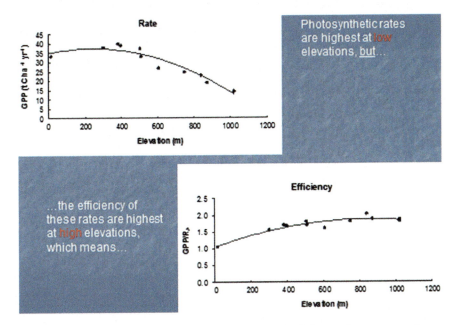

Fig. 7.16 Power terms showing inverse ratio of rate and efficiency of ecosysetms along a mountain elevational gradient, and a power curve with a maximum at intermediate rates

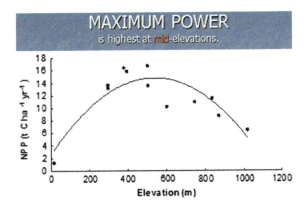

Fig. 7.17 Observed maximum power (net photosynthesis) is at intermediate elevations for forests on Luquillo Mountains, Puerto Rico

The maximum power principle addresses the empirical question of why systems of any type or size organize themselves into the patterns observed and assumes that physical laws govern system function. It does not assume, for example, that the system comprising economic production is driven by consumers; rather that the whole cycle of production and consumption is structured and driven by physical laws (Fig. 7.18). This does not mean that our actions are predetermined, or that humans are not free to choose "other life styles". Odum himself said that humans

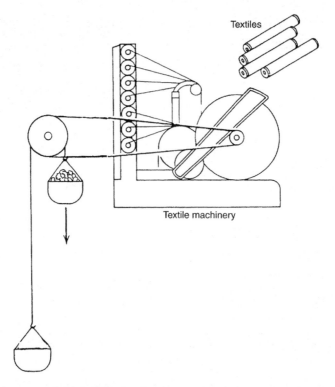

Fig. 7.18 A conceptualization of how maximum power is related to maximum production rate in an economic system, in this case by Atwood's machine being used to drive economic production machinery (From Hall 1995)

(as individuals or groups) were free to do what they wished, but that over time natural selection would insure that they would have to behave within energetic possibilities and selective advantages including, as social animals, within the norms of society. Those beliefs and social practices that do not align with the effects of the maximum power principle are unable to compete, and lose what social power base they may enjoy. In other words human behaviors are both driven by and constrained by energetic principles.

An extremely interesting, but not fully resolved, issue is the degree to which maximum power applies to explaining the behavior of nations in the exploitation of the world's resources. Like evolution itself, understanding this is not necessarily a warm and fuzzy experience. There appeared to be selection in the past for the great colonial exploitations – England and Spain in the Americas, England in India, France in Indochina, Japan in Asia, Incas and Mayans in the Americas – at least until growing populations and burgeoning nationalism in the colonies made the costs too great for the colonists. This is hardly new – my own reading of Plutarch's lives (of ancient Greeks and Romans) astonished me at how tyrants who defeated and sacked other cities were glorified by Plutarch. Whether whatever "Pax Romanas" were generated protected the subjugated people who might have been better off is another matter. The concept might apply now: if the US does not use Middle Eastern Oil does that mean China will be leaving us none? Are nations in competition to use up the world's resources at some kind of optimum rate, and will that undermine any hope for generating a world safe from climate change? Was there selection for neoclassical economics despite its many obvious logical flaws because it provided a means to accelerate the power of those societies that bought into it? Such questions are beyond the scope of this book, but may be critically important.

7.4 Proposals for Maximum Power Principle as the 4th Thermodynamic Law

There are many who believe that maximum power is a critically important but largely unappreciated principle (or perhaps law). One whose importance will become increasingly clear with time. Various people (e.g., Lotka 1922b) think it should be established as a fourth law of thermodynamics (the third states that the minimum energy of a system exists when the substance is in crystalline form). There are even some who suggest that the maximum power concept is a process that is inherent not just in thermodynamic systems, but also in highly abstracted systems such as modern social and financial systems and, as mentioned above, how global resources are exploited.

Taken to a logical extreme we must ask whether those beliefs, plans, religions, political theories, modes of governance, business models, or other social practices that do not align with the effects of the MPP cannot compete effectively against those that do. Natural selection reduces their physical base (biomass, adherents, social infrastructure, invested capital) until they fail and become extinct. On the other hand, those beliefs, etc., that do align with the effects of the MPP will see their physical base increase under the ministrations of the MPP. Is this what could explain the persistence of neoclassical economic theories? On the other hand Odum certainly believed in some kind of group (he would say systems) selection, and even admitted later in life that in times of decreasing energy availability (e.g., peak oil) there would be stronger selection for efficiency than power.

7.5 Conclusion

These three chapters show that biology, ecology, and natural selection can all be described elegantly and fully from an energy perspective based on the laws of thermodynamics and the principle of maximum power. They first shows how mechanisms of energy use enable physiological and behavioral activity. Then, they discuss how these mechanisms can be understood in terms of energy investments, EROI, and maximum power.

Acknowledgments I thank Garvin Boyle, Don Chisholm, Mary Odum-Logan and Jim Brown for assistance with this chapter.

References

Curzon, F.L., and B. Ahlborn. 1975. Efficiency of a carnot engine at maximum power output. *American Journal of Physics* 43: 22–24.

Gillman, L.N., S.D. Wright, J. Cusens, P.D. McBride Y. Malhi, and R.J. Whittaker. 2015. Latitude, productivity and species richness. *Global Ecology and Biogeography, (Global Ecol. Biogeogr.)* 24: 107–117.

Hall, C.A.S. 1988. An assessment of several of the historically most influential theoretical models used in ecology and of the data provided in their support. *Ecological Modeling* 43: 5–31.

Hall, C.A.S. 1995. Maximum power: the ideas and applications of H.T. Odum. University Press of Colorado.

Hall, C.A.S. 2000. *Quantifying sustainable development: The future of tropical Economies*. San Diego: Academic Press.

Harris, Nancy L., C.A.S. Hall, and A.E. Lugo. 2013. A test of the maximum power hypothesis along an elevational gradient in the Luquillo Mountains of Puerto Rico. *Ecological Bulletins* 54: 233–243.

Huston, M.A., and S. Wolverton. 2009. The global distribution of net primary production: resolving the paradox. *Ecological Monographs* 79: 343–377.

Langbein, W.A, and B. Leopold. 1966. *Theory of minimum variance*. Physiographic and hydraulic studies of rivers Geological Survey Professional Paper 422-H.

Lotka, A.J. 1922a. Contribution to the energetics of evolution. *Proceedings of National Academy of Sciences* 8: 147–151.

Lotka, A.J. 1922b. Natural selection as a physical principle. *Proceedings of National Academy of Sciences* 8: 151–154.

Odum, H.T. 1983. *Systems Ecology: An Introduction*. NY: John Wiley, 644 pp.

Odum, H.T. 1994. *Ecological and general systems: An introduction to systems ecology*. Niwot Colo: University Press of Colorado.

Odum, H.T. 1995. Self-organization and maximum empower, in C.A.S. Hall (ed.) *Maximum Power: The Ideas and Applications of H.T.Odum*. University Press of Colorado, Niwot Colo.

Odum, H.T., and R.C. Pinkerton. 1955. Time's speed regulator: The optimum efficiency for maximum output in physical and biological systems. *American Scientist* 43: 331–343.

Part III
Energy and Human Economies

This book has considered investments and energy in relation to their general properties and application to the basic physical and bilogical systems of the earth. We then introduced EROI with respect to biological systems in chapter 6, and maximum power in chapter 7. But all of these concepts apply just as fully to humans and their economies. The remaining chapters consider these concepts, and energy more generally, with respect to human economic systems.

Chapter 8
Energy in Early Human Economies

To chemists Frederick Soddy and William Ostwald, anthropologist Leslie White, archeologist Joseph Tainter, historian John Perlin, systems ecologist Howard T. Odum, sociologist Frederick Cottrell, economist Nicolas Georgescu-Roegen, energy scientist Vaclav Smil and a number of others in these and other disciplines, human history, including contemporary events, is essentially about the history of how humans have exploited energy and the technologies they have used to do so.

While they acknowledge that other issues, including human culture, water, nutrient cycling, and entropy (among many others) can be important, thay all share the opinion that it is energy itself, and especially surplus energy, which is the key. Survival, military efficacy, wealth, art, and even civilization itself are believed by all of the above to be a product of surplus energy. For these authors the issue is not simply whether there is surplus energy but how much, what kind (quality), at what rate it is delivered and how it is invested. The interplay of these factors determined net energy and hence the ability of a given society to divert attention from life-sustaining needs such as agriculture or the attainment of water towards luxuries such as art and scholarship. Indeed humans could not possibly have made it this far through evolutionary time, or even from one generation to the next, without there being some kind of net positive energy, and they could not have constructed such comprehensive cities, civilizations, or wasted so much in war without there being substantial surplus energy in the past.

8.1 Application to Our Early Ancestors

"Not the slightest scrap of hard evidence, either morphological or genetic, exists to suggest that Homo sapiens is not, like all other animals, a natural product of evolution. Therefore we, like they, are uncontaminated by supernatural influences, good, bad, or divine. We may well be excellent communicators and toolmakers, but overwhelming evidence shows that these distinctions are of degree, not of kind. The only irrefutable argument in favor of humanity's specialness is in fact purely mystical—and entirely circular." (Morrison 1999).

So humans, like the plants and animals discussed in the previous three chapters, must generate sufficient net energy to survive, reproduce, and adapt to changing conditions. Humans are different from other species in that we have the ability to evolve culturally at rates far greater than conventional organic evolution. As a consequence of this cultural evolution human numbers have increased dramatically since their emergence as a species in Africa half a million years ago, and especially in the last several hundred years. Again I argue that the most important aspects of this evolution have been energy-based. A key evolutionary event was the development of spear heads and knife blades, which are energy or force concentrating devices that allowed humans to exploit a much broader and larger animal-resource base for food and skins. Another important event was the harnessing of the energy contained in the chemical bonds of wood, using fire, which allowed humans to exploit even more food resources by breaking down plant cell walls through cooking, and to smelt metals, kiln cement, and bake ceramics (Wrangham 2009). All these developments assisted humans in their exploitation of colder, more northerly ecosystems. The two most important of these new energy-based technologies were agriculture, which redirected an area's photosynthetic energy from natural to human food chains, and the learning over the past two centuries of how to exploit the enormous quantity of concentrated energy in the chemical bonds of fossil (meaning old) fuels, principally coal, oil, and natural gas.

8.2 Hunter-Gatherers

Human populations must generate sufficient net energy to survive, reproduce, and adapt to changing conditions. They must first feed themselves before attending to other issues. While people in most industrial societies today hardly worry about getting enough to eat, for much of the world and much of humanity's history and prehistory getting enough food was the most important issue. For at least 98 % of the 2 or so million years that we have been recognizably human, the principal method by which we as humans have fed ourselves, that is obtained the energy we need for life, has been that of hunting and gathering. In this case humans invest their own energy into getting more energy—both in terms of food energy. Contemporary hunter-gatherers—such as the !Kung of the Kalahari desert in southern Africa (portrayed sympathetically and perhaps romantically in the movie "The gods must be crazy")—are probably as close to our long term ancestors as we will be able to understand. Most hunter-gatherer humans were probably little different from cheetahs or trout in that their principal economic focus was on obtaining food and securing surplus energy directly from their environment. Most studies by anthropologists such as Lee (1969) and Rapaport (1967) confirmed that indeed present-day (or at least recent) hunter-gatherers and shifting cultivators acted in ways that appeared to maximize their own energy return on investment. This is also found in more recent studies (e.g., Glaub and Hall, in revision).

An especially appropriate study seems to be that of Lee. According to Lee the !Kung life style, under normal circumstances, generates a quite positive energy return on investment (i.e., generates a large surplus) from their desert environment, perhaps an average of some 10 kcal food returned per their own kcal invested in hunting and gathering. In normal times these cultures had plenty to eat, and the people tended to use the time made available from their relatively high EROI lifestyles in socializing, childcare, and story-telling. The downside was that there were periodic tough times, such as droughts, during which starvation was a possibility. It is probable that the lifestyle of all our ancestors had a fairly positive EROI for much of the time, although periodic droughts, diseases, and wars occasionally, or perhaps routinely, took a large toll. Thus even though they had a relatively high EROI, perhaps 10:1, their populations tended to be relatively stable over a very long time, for human populations barely grew from thousands of years BC until about 1800. Thus even a relatively high energy return was not enough to generate much in the way of net population growth over time. Long-term breast feeding probably helped space out children.

The rate at which most plants and animals can exploit their own resource base changes only very slowly through evolution. All must adapt to climate and other changes by selective gene shifts or occasional mutations, and animals must also adapt to the fact that their food is also going through its own defensive evolutionary changes. Humans are different, for the human brain, language and the written word have allowed for much more rapid cultural evolution. The most important of these changes were energy-related technologies: the development of energy (or force)-concentrating spear points and knife blades, agriculture as a means of concentrating solar energy for human use, and more recently the exploitation of wind and water power and fossil fuels. Geneticists talk of "memes" versus genes, the former being culturally forwarded information of utility to survival and reproduction—sometimes called cultural evolution. The process is far more rapid and effective, at least in the short term, than organic or genetic evolution.

8.3 Agriculture and Deforestation

Each of these cultural adaptations is part of a continuum in which humans increase the rate at which they exploit additional resources, both energy and otherwise, from nature. The development of agriculture allowed the redirection of the photosynthetic energy captured on the land from the many diverse species in a natural ecosystem to the few species of plants (called cultivars) that humans can and wish to eat, or to the grazing animals that humans controlled. Curiously the massive increase in food production per unit of land brought on by agriculture did not, on average, increase human nutrition over the long run but mostly just increased the numbers of people (Angel 1975). Of course it also allowed the development of cities, bureaucracies, hierarchies, the arts, more potent warfare, and so on—that is, all that we call civilization, as nicely developed by Jared Diamond in his famous book Guns, Germs and Steel (Diamond 1998). Throughout most of human history, humans themselves did most of the physical work, often as slaves but more generally as physical laborers which, in one way or another, most humans were.

Over time humans increased their control of energy through technology. For thousands of years most of the energy used was animate—people or draft animals—and derived from recent solar energy. A second very important source of energy was from wood, which has been recounted in fascinating detail in Redman (1999), Perlin (1989), Pointing (1992) and Sml (1994). Massive areas of the Earth's surface —Peloponnesia, India, parts of England and many others have been more or less completely deforested three or more times as civilizations have cut down the trees for fuel or materials, prospered from the newly cleared agricultural land and then collapsed as fuel and soil became depleted. Archeologist Joseph Tainter (1998) recounts the general tendency of humans to build up civilizations of increasing reach and infrastructure that eventually exceeded the energy available to that society. Thus from megafaunal destruction by paleo-hunters to ancient agriculture-based civilizations human degradation and destruction of the environment is hardly a new phenomenon (Martin 1973; Redman 1999).

8.4 Were Early Human Societies Sustainable?

Were humans ever truly sustainable? As hunter-gatherers they certainly had low impact on their environment, although even that is debatable because of their apparent contribution to "megafaunal extinction" of large animals (Martin 1973). The development of agriculture almost necessarily leads to the slow depletion and eventual exhaustion of the fuels and soils used as well as many other types of environmental degradation. Agronomist David Pimentel summarized it as "agriculture will work until we have titrated out our soils and our fossil fuels, a matter of at most a few hundred more years." Fossil fuels certainly have generated an enormous increase in the ability of humans to do all kinds of economic work, greatly enhancing what they once did with their own muscles or those of work animals. This work includes the production of food, and the increase in food (aided by increased knowledge of basic sanitation and public health) has led to the enormous increase in the number of humans. Since human societies are still here they were, by definition, sustainable. Whether our modern industrial society is sustainable is quite another matter, considered in the last chapter.

8.5 Expense of Energy to Early Civilizations

There is relatively little quantitative information about actual EROIs for energy producing systems from the medium or distant past. Sundberg (1992) made a quite detailed assessment of the energy cost of energy in earlier Sweden. From 1560 until 1720 Sweden was the most powerful country in Northern Europe, based mostly on its very productive metal mines, but also an aggressive foreign policy backed up by high-quality metal weapons. The production of these mines required enormous amounts of energy for mining and especially smelting. The source of this energy was wood and especially charcoal (needed to get the high temperatures steel required) cut from Swedish forests. Sundberg gives a detailed calculation of how a typical forester and his family, self-sufficient on 2 ha of farmland, 8 ha of pastures and 40 ha of forest (collectively intercepting 1500 TJ of sunlight) generated some 760 GJ of charcoal in a year for the metal industry (equivalent in energy to 125 barrels of oil). To do that required about half a GJ of human energy or 3.5 GJ if we include the draft animal labor. So we might calculate the EROI of the human investment to be as high as 1500:1, or some 220:1 if we include the animals. But that is just the direct energy, as it took 105 GJ to feed, warm, and support the forester and his family (which includes his replacement) and probably at least that to support the animals. So if we include direct plus indirect energy the EROI is down to roughly 7:1. The system was sustainable as long as the forests were not overharvested. That was true until the middle of the nineteenth century, but then the forests were severely overharvested and many Swedes left for America.

8.6 EROI Analyses Over Very Long Periods of Time

There have been several attempts to derive EROI or similar analyses over long time periods, by assuming that energy costs are related to financial costs. Court and Fizaine (in revision) examined the EROI of global fossil fuels from 1800 to the present and found that EROI was usually about 30-40:1 for all fuels until 1920 but increased to about 60:1 for oil amd coal, and more than 100:1 for gas during the 1960s. All declined precipitously during the 1970s, then increased again until the 1990s, in mirror image of prices as has been found in other studies. There was some evidence that the EROI may be decreasing again after the secondary high point of the 1990s. Oil and gas have subsequently declined irregularly to 10:1 and 20-40:1 respectively, and coal increased to about 80:1, in more recent years. As mentioned King et al. (2016) analyzed the cost of all energy (which might be considered a reciprocal of EROI; see King and Hall 2011), including wood, fodder and food and later other types from 1300 to the present. He found that for most of this time energy costs were roughly 40 % of GDP (implying an EROI of 2.5:1, assuming that the surplus energy ran the rest of society), but that with the industrial revolution the costs began to decline until they reached about 5–10 % of GDP.

References

Angel, J.L. Paleoecology, paleodemography and health. In *Population ecology and social evolution*, ed. Polgar S., 667–679. Mouton: The Hague, 1975.

Court, V., and F. Fizaine, in revision, Long-term estimates of the global energy-return-on-investment (EROI) of coal, oil, and gas. *Ecological Economics*. (In revision).

Diamond, J. 1998. *Guns, germs and steel*. Norton, N.Y.

King, C., and C.A.S. Hall. 2011. Relating financial and energy return on investment: sustainability. Special Issue on EROI. 1810–1832.

King, C.W., John P. Maxwell, and Alyssa Donovan. 2015. Comparing world economic and net energy metrics, Part 1: Single technology and commodity perspective. *Energies* 8: 12949–12974.

Glaub, M., and C.A.S. Hall, in review, EROI of subsistence hunting by !Kung tribesmen. (manuscript in revision for Human Ecology).

Lee, R. 1969. !Kung bushmen subsistence: an input-output analysis, pp 47–79. In *Environment and cultural behavior; ecological studies in cultural anthropology*, ed. Vayda, A.P. Published for American Museum of Natural History [by] Natural History Press: Garden City, N.Y.

Rappaport, Roy A. 1967. *Pigs for the Ancestors*. New Haven: Yale University Press.

Morrison, R. 1999. The spirit in the gene: Humanity's Proud Illusion and the Laws of Nature. Cornell Univ. Press, Ithaca N.Y.

Martin, P.S. 1973. The discovery of America. *Science* 179: 969–974.

Redman, C.L. 1999. *Human impact on ancient Environments*. Tucson: University of Arizona Press.

Sundberg, U. 1992. Ecological economics of the Swedish baltic empire: an essay on energy and power, 1560–1720. *Ecological Economics* 5: 51–72.

Tainter, J.A. 1988. *The collapse of complex societies*. New York: Cambridge University Press, Cambridge Cambridgeshire.

Perlin, J. 1989. *A forest journey: the role of wood in the development of civilization*. New: W.W. Norton.

Pointing, C. 1992. A green history of the world: the environment and the collapse of great civilizations. Penguin,.

Smil, V. 1994. *Energy in world history*. Boulder: Westview Press.

Wrangham, R. 2009. Catching fire: how cooking made us human. Basic Books. N. Y.

Chapter 9
Fossil Fuels

The principal energy sources in antiquity were all derived directly from the Sun: human and animal muscle power, wood, flowing water, and wind. About 300 years ago the industrial revolution began. It brought an exponential increase in the energy available to humans to do economic work. This revolution began with stationary wind-powered and water-powered technologies, which were subsequently supplemented and replaced by fossil hydrocarbons (fossil meaning old): coal in the nineteenth century, oil since the twentieth century, and now, increasingly, natural gas. The global use of hydrocarbons for fuel by humans has increased nearly 800-fold since 1750 and about 12-fold in the twentieth century. The enormous expansion of the human population and the economies of most nations in the past 100 years have been facilitated by a commensurate expansion in the use of fossil fuels (Fig. 9.1).

The industrial revolution began on a small scale about 1750 but then increased rapidly for the next two and a half centuries. During this time there has been a remarkable change in the quantity and quality of hydrocarbons humans used, from the comparatively dilute forms of recently captured solar energy of wood and muscle power to the vastly more concentrated fossil fuels. This was the beginning of the modern world. Humans had begun to understand how to exploit the much more concentrated energy found in fossil fuels. Fossil hydrocarbons have greater energy density than the carbohydrates such as food and wood, and as a consequence they can do much more work—heat things faster and to a higher temperature, operate machines that are faster and more powerful and so on (Table 9.1).

While some have argued that we live in an information age, or a post industrial age, both contentions are only barely true. Overwhelmingly we live in a petroleum age. Just look around. Transportation, food production, plastics, most of our jobs and leisure, much of our electricity, and all of our electronic devices are dependent upon gaseous and especially liquid petroleum. This has been, and continues to be, the age of oil, and of hydrocarbons more generally. Perhaps the industrial revolution should be renamed "the hydrocarbon revolution" because that is what

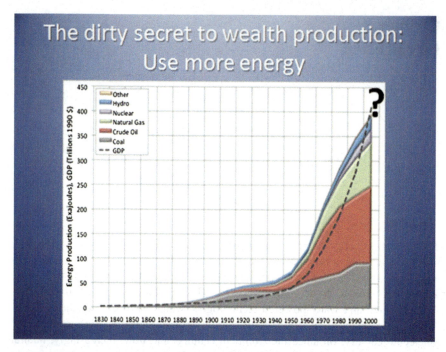

Fig. 9.1 To many energy analysts the expansion of cheap fuel energy has been far more important than business acumen, economic policy or ideology in this expansion of wealth, although they too may be important. The question mark implies that the future growth of both energy use and economies is very uncertain, each of which has slowed down some 2010–2016 (Image courtesy of David Murphy)

Table 9.1 Evolution of power outputs of machines available to humans

Machine	Horsepower	kW
Man pushing a lever	0.05	0.04
Ox pulling a load	0.5	0.4
Water wheels	0.5–5	0.4–3.7
Versailles water works (1600)	75	56
Newcomen steam engine	5.5	4.1
Watt's steam engine	40	30
Marine steam engine (1850)	1000	746
Marine steam engine (1900)	8000	6000
Steam turbine (1940s)	300,000	22,400
Coal or nuclear power plant (1970s)	1,500,000	1,120,000

Derived from Cook (1976). Man, Energy, Society, W. H. Freeman happened—humans have moved from using various carbohydrates to using hydrocarbons as their principle fuel for doing economic work.

Why did they do this? The answer is simple. People wanted to do more work because that made them more profits, and fossil fuels generated more energy and hence financial return on their investments. They wanted more of some raw material transformed into something useful that they can eat, enjoy, trade, or sell. Fossil fuels are a remarkably cheap way of fueling this additional work. To illustrate: a strong human being might provide 500 MJ of work over a year for wages of (in US) perhaps 30 thousand dollars. But a single barrel of oil contains ten times that amount of energy and costs but 50 dollars. Each would work at roughly 25 % efficiency. So hundreds to thousands of times more work per dollar can be achieved from fossil fuels compared to human workers.

The initial push in the development of modern civilization came with learning how to burn coal to do economic work with the energy released, especially making iron and running steam locomotives. The use of coal to do many things was largely enabled by the invention of the Watt steam engine between 1763 and 1775. By 1860, Stanley Jevons was able to say that there was hardly any economic activity in England that did not lead back to coal. Over the last century humanity overwhelmingly transitioned from the age of coal to the age of oil, or if we want to include natural gas, the age of petroleum.

It was not simply the development of the use of coal. A whole suite of related financial, chemical, metallurgical, and other developments took place around the same time which accelerated each other and led to the enormous production of wealth that took place in England, Scotland and then other European countries during the 1800s. For example, James Watt could not develop his famous steam engine until his friend William Wilkinson had perfected the iron refining and drilling technologies that allowed for the construction of a perfectly round cylinder. Even their interactions required the social environment of the Scottish enlightenment for their ideas to evolve and to come to fruition as actual components of society. Most thinking people at that time believed that these were wonderful inventions that would finally free people from the drudgery of everyday activities and allow them to build a better society through rational thinking. At the same time many of the English Romantic poets, notably William Wordsworth, were horrified by the smoke and grime and repetitive jobs of the industrial revolution and pined for the bucolic preindustrial England.

Our societies today need vast amounts of energy mined from stocks of solar energy accumulated over eons, and converted into coal, natural gas, and petroleum. Without these stocks we could not live as we do. Clearly the world now has at its disposal a tremendous amount of power compared to the past (Figs. 9.2 and 9.3). Some popular and some scientific presses suggest that we have entered a "postindustrial" society, where computers and, more generally, human knowledge have replaced raw energy and materials in the generation of wealth. This is easy to say while each of us has the equivalent of some 60–80 energy slaves "hewing our wood and hauling our water." Neither knowledge nor computers can do physical work on their own. The raw fact is that the enormous increase in human numbers and their affluence is a consequence of our exploitation of fossil fuels.

Fig. 9.2 33 Horsepower harvester. Controlled by 5 workers. In addition other workers were needed to till the land for horse food, water and feed the horses, muck the stables, etc. Photo courtesy of Mario Giampietro

9.1 Economic Implications of Fossil Fuels

To many energy analysts that expansion of cheap fossil fuel energy has been far more important than business acumen, economic policy, or ideology in this expansion of wealth, although they too may be important (Fig. 9.1). While we are used to thinking about the economy in monetary terms, those of us trained in the natural sciences consider it equally valid to think about the economy and economics from the perspective of the energy required to make it run. When one spends a dollar, we do not think just about the dollar bill leaving our wallet and passing to someone else's. Rather, we think that to enable that transaction, that is to generate the good or service being purchased, an average of about 6 MJ of energy (roughly equivalent to the amount of oil that would fill half a standard coffee cup) must be extracted and turned into roughly a half kilogram of carbon dioxide. Take the money out of the economy and it could continue to function, at least in theory, through barter, albeit in an extremely awkward, limited, and inefficient way. Take the energy out and the economy would immediately contract or stop completely. Cuba found this out in 1991 when the Soviet Union, facing its own oil production and political problems, cut off Cuba's subsidized oil supply. Both Cuba's energy use and its GDP declined immediately by about one third. Groceries disappeared

9.1 Economic Implications of Fossil Fuels

Fig. 9.3 200 horsepower fossil-fueled harvester, controlled by one person and not requiring fuel in the off-season. Other workers are required to make and repair the machine, and to provide the diesel fuel. Photo courtesy of Mario Giampietro

from market shelves within a week and soon the average Cuban lost 20 pounds. Cuba subsequently learned to live, in some ways well, on about half the oil as previously, but the impacts were enormous and extremely challenging.

9.2 Efficiency in Energy Use

Many argue that through technology and markets we are becoming more efficient in generating wealth per unit of fuels used, and there is much evidence for that. For example since 1980 the energy use of new homes per square foot has declined by nearly 20 %. The industrial energy use per unit value of product is down by nearly 40 % and the fuel economy of passenger vehicles has improved by more than 25 %. Overall the United States has gone from using about 13.5 MJ per 2016 dollar of GDP in 1970 to 5.5 MJ per dollar, a remarkable increase in efficiency (Nadel et al. 2015). But while this would seem to be unequivocal, a closer examination shows a bit more ambiguity. Between 1970 and 2009, the average size of a new home in the US grew from 1400 to 2700 square feet, more than offsetting the efficiency gains. Energy analyst Robert Kaufmann suggests that while there has been some real improvements in fuel efficiency (driven mostly by higher fossil fuel prices) the increases in efficiency are due principally to a shift to higher quality fuels, and especially to structural changes in national economies as richer nations moved their heavy industries overseas to reduce pollution or find cheaper labor (Kaufmann 2004). There may be another reason as well that the United States, but few other nations, appears to be becoming more efficient in our use of energy. According to the (controversial) organization Shadow stats, the U.S. has been engaged in a systematic "cooking of the books" on the official measure of inflation, that is a deliberate official underestimate of inflation since 1985 to make governments look good. Correcting for any or all of these actions would greatly decrease the perceived improvements of efficiency in the US economy.

Likewise in other countries the evidence is not quite clear (e.g. Hall and Ko 2005; Tharaken et al. 2001). Smil (2007) reports that over a recent decade the energy efficiency of the Japanese economy had actually decreased by 10 %. Further Ayres et al. (2014) found that most improvements in "technology" in Europe have been simply an increase in the quantity of energy used or improvements in getting the energy to the point where the work is done. Although neoclassical economic models purport to show that technology alone has driven the increase in production of the industrial economy, historically, it has been a technology that mostly has found new sources of, and applications for, energy.

Further compounding our understanding of the efficacy of efficiency is what is called "Jevon's Paradox", or, sometimes, the rebound effect. Stanley Jevons, considered to be perhaps the most generally knowledgeable engineer in England in 1850, was asked by Queen Victoria's advisors to study "coal." To make a long story short Jevons initially recommended that given the importance of coal to the British history and the limited quantities remaining relative to Britain's long history, there was a need to improve its efficiency of use. But then he found three earlier studies that had also recommended that, and that each time "Every improvement [in the efficiency] of the [steam] engine, when effected, does but accelerate anew the consumption of coal" Jevons (1865). This has also been seen in other ways: when American cars were made more efficient from 1970 to 2000 people drove them more miles, when refrigerators were made more efficient people bought larger ones with more gadgets, when houses became better insulated people bought larger houses and so on. As a society the U.S. has seemingly became more efficient, but there are many reasons for that, including exporting the manufacture of energy intensive things we use and possibly a misspecification of inflation rates, debts and so on (see above paragraphs). Meanwhile overall the U.S. continues to use the same or more energy, not less.

9.3 Fossil Energy and Economics

Economics is mostly studied and taught as a social science, with very little connection to the natural sciences except for (1) occasional and sometimes disastrously inappropriate excursions into the emulation of the mathematical sophistication (but not, alas, the reality mooring) of physics and (2) attempts to measure the value of nature in economic terms. This book has developed the reality that economies are in many ways completely dependent upon energy for their operation and indeed are basically about how energy is used to transform raw materials from nature into the products and services that are traded in markets. Economics as a discipline should reflect this basic reality, but essentially does not. One explanation is the training of most economists in the social sciences rather than natural sciences. But this problem is not entirely the fault of economists themselves, for the teaching of the natural sciences also is too often divorced from its application to the day to day issues of interest to economists.

Modern neoclassical economics, the dominant form of economics taught and practiced today, is grounded in social sciences and especially "consumer sovereignty." Consumer sovereignty is the proposition that value, like beauty, is in the eye of its beholder—that is a matter of *subjective* well-being or utility. The overall objective of these economists tends to be not to understand the origins of wealth (as was the focus of some earlier economists), as much as to show, under ideal theoretical conditions, that market economies are self-regulating by means of small, or marginal, fluctuations in prices driven by competition on the individual level. The result of voluntary trades, based solely on the maximization of self interest, (in theory) leads to a situation of *Pareto efficiency* (named after its originator, Vilfredo Pareto) where no one individual can be made better off without making another worse off. Under these assumptions government intervention could do no good, and much harm, as it would distort the signals of the market, which is seen as a perfect carrier of information. Careful objective analysis, for example by Hall et al. (2001), Sekera (2016) and many others, or logical analysis (e.g., by Leontief 1982), has shown this not to be the case, but that seems to be of no interest to most economists, who love their theory.

Essentially all contemporary economists are oriented to greater or lesser degree toward the continuation and enhancement of economic growth. Their perspective is that in the absence of growth employment would stagnate and human well-being would decline. Meanwhile the concept that resource shortages might be important is simply not considered.

9.4 The Role of BioPhysical Economics

What, then, should be the purpose of BioPhysical economics, the approach we are advocating in this book? Clearly we must deal with a world that is increasingly dependent upon stocks of fossil fuels, the depletion of those stocks, the high energy costs of substitutes and the increasing difficulty of achieving growth as depletion occurs. Unlike the utilitarians, BioPhysical economics considers and encourages the possibility that humans are capable of achieving happiness by means other than the acquisition of ever-increasing quantities of material goods—goods that cannot be produced with declining resources. As such it calls back to the center stage the question of distribution: for generations that question has been suppressed for if the pie has been getting larger then everyone can get a larger piece. But if the size of the pie is not growing, who should get how large a piece?

It is not the purpose of this book to derive BioPhysical economics in any detail. That is done in, e.g., Hall and Klitgaard (2012). But in a certain way BioPhysical economics is the logical conclusion of all of the energy facts, logic, and arguments that we have presented here, at least with respect to the human condition, and as such deserves careful consideration by the reader.

The suggestion to limit human economic activity for environmental reasons is not new. BioPhysical economics serves as a wakeup call to the impending and inevitable end of the economy based on high quality fossil fuels, and with it the probable end of growth economics. It also provides important caveats as to which of the many alternatives proffered has a good chance of succeeding by providing guidelines for the assessment of alternative sources of energy. How we can live well within nature's limits is a question we can no longer afford to postpone or subsume to a series of equations unconstrained by reality. To answer this whole new set of questions we must first assess how economists have addressed the age-old ones, for these questions remain as relevant for these new conditions as they were for the circumstances when they were asked. In other words, for only a relatively few decades—a century and a half at most—in the most favorable situations has a year by year increase in general affluence been the normal condition. It was not true back when early economists were writing and it appears no longer true for much of the world. So we must pay attention once again to their questions—but we need to do that while including an energy perspective and a consideration of either limiting growth or having it done for us by nature.

In 1971 biologist Paul Ehrlich and physicist John Holdren proposed the environmental IPAT equation, which says that environmental Impact equals Population level times per capita Affluence times a Technological factor (which could be positive or negative). To many of us this was a pretty good summary of where environmental impact comes from. But population is an issue that we have been unwilling to talk about—let alone do anything about—and since that time has disappeared from our public view. Reducing affluence to reduce environmental impact is even less popular, and amounts to political suicide. Most governments of the world are doing more or less what they can to get back to the higher economic growth of even a decade ago. Thus society has focused on only the technical factors which are woefully unequal to the task. If energy resources are indeed as restricted as many analysts think, then we have some serious new thinking to do.

Finally, from the perspective of this book, we have to ask some new questions about investments. Past investments—over the past century—were made at a time when the production of high quality fossil fuels was increasing at rates as high as 5 % a year. At the time of this writing they have declined to no more than 1 % a year (Fig. 9.4a), and the US (and global) economies show similar pattern (Fig. 9.4b). There are many projections that the production of all fossil fuels will peak within 10 or at most 50 years (Fig. 9.5). If this is the case then there will be

9.4 The Role of Biophysical Economics

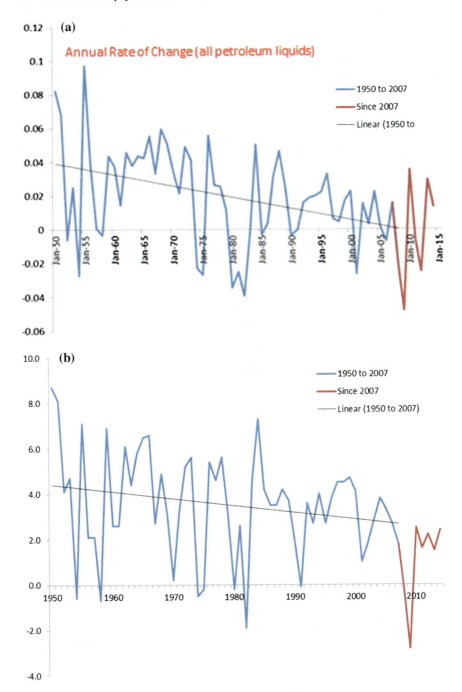

Fig. 9.4 **a** Growth rate of global oil production: 1950–2015. **b** Growth rate of U.S. Economy: 1950–2015. Growth rates in Europe, Japan and many other "modern" countries are similarly declining but lower

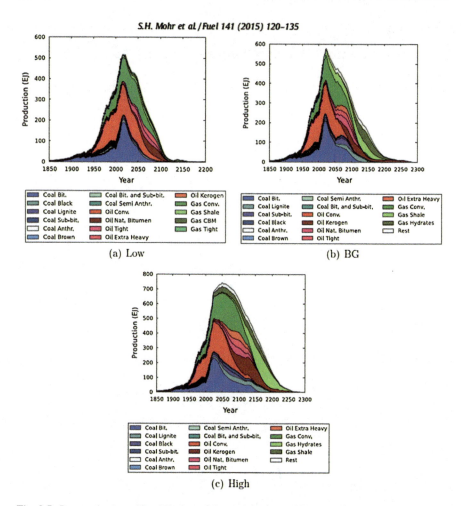

Fig. 9.5 Past production of fossil fuels, and three projections of future availability: **a** low estimate **b** best guess and **3** high (maximum) estimate (from Mohr et al. 2015). Similar estimates are available from Maggio and Cacciola (2012); and from Laherrere (France ASPO web site). All three estimates are showing a peak of all fossil fuels within 10–35 years at most

many economic pressures to maintain consumption. If fuel supplies decrease while pressures to maintain consumption does not, it implies that there will be little left over to make the investments needed to maintain production or shift it to something else. Heinberg and Fridley (2016) estimate that if we are to shift to a 100 % solar society over the next several decades the financial investments required will need to be 20 times all of our renewable investments at the present in each of 20 years. The energy investments would be of a similar magnitude. This will certainly be the largest issue ever to confront humans: how to generate enormous new energy investments while energy and EROI are declining.

References

Ayres, R.U., J.C.J.M. Van den Bergh, D. Lindenberger and B. Warr. 2013. The underestimated contribution of energy to economic growth. Structural Change and Economic Dynamics 27, 79–88.

Cook, E. 1976. Man, Energy, Society, W. H. Freeman. San Francisco.

Hall, C.A.S. and J.-Y. Ko. 2005. The myth of efficiency through market economics: a biophysical analysis of tropical economies, especially with respect to energy, forest and water. pp. 40–58, in Bonell, M., and L.A. Bruijnzeel (Eds.), *Forests, Water and People in the Humid Tropics: Past, Present and Future Hydrological Research for Integrated Land and Water Management.* Cambridge University Press. Cambridge, UK. 944 pp.

Hall, C.A.S., D. Lindenberger, R. Kummel, T. Kroeger, and W. Eichhorn. 2001. The need to reintegrate the natural sciences with economics. *BioScience* 51: 663–673.

Hall, C., and K. Kent 2012. Energy and the Wealth of Nations. *Understanding the Biophysical Economy.* Springer, N.Y.

Heinberg, R. and D. Fridley 2016. Our renewable future: laying the path for one hundred percent clean energy. Island Press, Washington, D.C.

International Energy Agency. 2016. *Fossil fuel subsidy database.* Available online: http://www.worldenergyoutlook.org/resources/energysubsidies/fossilfuelsubsidydatabase/. Accessed on 10 August 2016.

Jevons, W.S. 1865. The coal question: An inquiry concerning the progress of the nation, and the probable exhaustion of our coal mines, in Flux, A.W. 1965. *Reprints of economic classics.* New York: Augustus M. Kelley.

Kaufmann, R.K. 2004. The mechanisms for autonomous increases in energy efficiency: a cointegration analysis of the US energy/GDP ratio. *Energy Journal* 25: 63–86.

Leontief, W.W. 1982. Academic Economics. *Science* 217: 104–107.

Maggio, G., and G. Cacciola. 2012. When will oil, natural gas, and coal peak? *Fuel* 31(98): 111–123.

Mohr, S.H., J. Wang, G. Ellem, J. Ward, and D. Giurco. 2015. Projection of world fossil fuels by country. *Fuel* 1(141): 120–135.

Nadel, S., N. Elliott, and T. Langer. 2015. *Energy efficiency in the United States: 35 years and counting report E1502 American council for an energy-efficient economy.* Washington, DC.

Sekera, J.A. 2016. *The public economy in crisis: A call for a new public economics.* Springer Briefs in Economics. NY.

Smil, V. 2007. Light behind the fall: Japan's electricity consumption, the environment, and economic growth. *Japan Focus,* April 2.

Tharakan, P.J., T. Kroeger, and C.A.S. Hall. 2001. Twenty-five years of industrial development: a study of resource use rates and macro-efficiency indicators for five Asian countries. *Environmental Science & Policy* 4: 319–332.

Chapter 10
EROI and Industrial Economies

The concept of EROI was introduced in Chap. 6 and applied there to issues of biology and evolution. But the concept is pervasive, and certainly applies very much to modern industrial economies. The following three chapters introduce the concept, show how it is derived and applies it to the American and world economies.

10.1 Introduction

EROI stands for Energy Return On Investment, and refers most explicitly to the ratio of energy delivered to an organism or society from one energy unit invested in getting that particular energy. The units can be Joules per Joule or barrels (of oil) per barrel, etc., and there can be modifications for, e.g., the quality of the energy input and output. It is a physical concept, but one that can have enormous economic implications, and one that must eventually be a large component or even determinant of many economic trends and assessments—for example, the price of oil (King and Hall 2011). Although EROI (or its conceptual cousins net energy analysis and life cycle analysis) is a term that is rarely heard today we believe that it is likely that this issue will become a dominant one in the U.S. and the world in coming decades because of the apparent substantial and continuing decline in EROI for the most important fuels and because alternatives (i.e., substitutes) tend to have a much lower ratio. The situation can be seen clearly for the finding and production of domestic oil in the United States. The EROI has evolved from the situation in 1970 when we found 25–35 barrels (or its equivalent as natural gas) for every barrel invested to 11 to 18 barrels per barrel in the 1990s (Cleveland et al. 1984; Cleveland 2005; Guilford et al. 2011) to less than 10 barrels per barrel today. The decline is even more striking if we examine the ratio of new oil found per unit invested, which has declined from more than 1000:1 in 1930 to 5:1 today (Guilford et al. 2011). The very large difference between the investment and the return, that is

© The Author(s) 2017
C.A.S. Hall, *Energy Return on Investment*, Lecture Notes in Energy 36,
DOI 10.1007/978-3-319-47821-0_10

the net energy surplus, of the oil industry allowed Texas and the United States to generate enormous wealth over the twentieth century. Initially, the quantity of oil produced increased dramatically each year from when it was first produced until its peak production in 1970. The EROI followed a somewhat similar path, increasing in the first half of the twentieth century then decreasing more or less routinely since 1970 for both extraction and especially for new discoveries. I believe that if the EROI for our principal fuels continues to decline the implications will be enormous as more and more of our total energy output, and hence our total economic activity, is diverted to get the same quantity of fuels (Fig. 10.1).

Within the scientific community EROI issues have had the greatest visibility when there have been rancorous debates about particular numbers. These debates are whether EROI is positive or negative for corn-based ethanol, by law some 10 % of the gasoline purchased in the United States, and what the EROI is for photovoltaic (solar) electricity. These issues will be covered in the next chapter, although I wish to point out that I believe that these issues, while valid, miss some of the richness and importance of how we might think about EROI, and can lead to erroneous policy decisions. Therefore, the objective of this and the next two chapters is to review and discuss the procedures by which we calculate EROI, summarize some pertinent literature values of EROI, identify what some of the reasons for existing discrepancies are and assess the implications of EROI for our future energy and economic situation, assess where it generates useful information that may be missed by market analyses, and where it may be less useful.

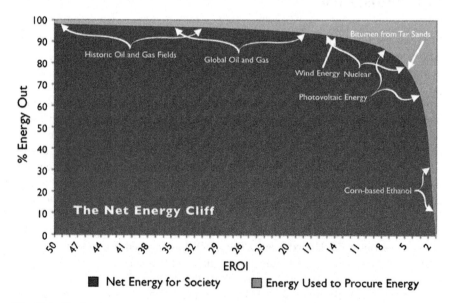

Fig. 10.1 The "net energy cliff." As EROI declines from high values to lower values it make little difference to society, as the net energy delivered does not change much. But below an EROI of 10:1 any declines makes a large difference on the amount delivered (*darker color*) (Courtesy of Euan Mearns)

10.2 Why Should EROI Change Over Time: Technology Versus Depletion

There are two basic perspectives, with two very different groups of followers, relating to the long-term trend of efficiency in the production of oil (and other nonrenewable) resources. Many resource analysts emphasize the importance of depletion as humans exploit and eventually exhaust higher grades (i.e., more concentrated resources from more accessible deposits) over time. On the other hand many economists and others have emphasized the importance of technology as a factor that either compensates for the decline of resource grades over time or that increases the production or utility of a resource, or the economy in general. This perspective was strongly cemented by the very influential paper by Barnett and Morse (1963) who found no indication of increasing scarcity of 13 mineral and several non-mineral categories of raw materials (except for forest products), as determined by their inflation–corrected price, for the first half of the twentieth century. Their analysis is still routinely trotted forth as evidence that there has been no increase in scarcity, or at least that technology had compensated for whatever increase in scarcity had occurred.

On the other hand many resource analysts have expressed serious concern about the depletion of fuels and other resources over time. Why should this happen? Well essentially any business, such as a mining enterprise, is interested in making the maximum profit possible. I start with copper, essential for many things including electricity generation and distribution. As of 1900 known copper ores in the United States ranged from roughly 40 % copper (in Butte, Montana) down to 4–0.4 %. The average grade mined was perhaps 4 % then, but has declined to about 0.4 % today (Lovering 1969). The high-grade ore is long gone. Without huge monetary and energy investments we would have little or no new copper today. Other things being equal it takes about the same amount of effort to dig up a ton of low-grade ore as a ton of high-grade ore, but there would be more salable product from the high-grade ore and hence more profit for the same effort to get the ton of high-grade ore. Thus there is a large incentive to use the highest remaining grade of essentially any resource first, and it would make no sense to use the lower grade as the cost would be something like ten times more per unit of salable copper. The *best first* principle states that humans use the highest quality sources of natural resources first as this would lead to higher profits. This concept was also of great interest to the classical economist David Ricardo. Given a choice, humans will grow crops on the more fertile soils, mine copper that is 10 rather than 1000 ft. deep, harvest timber from forests that are closer to roads and sawmills, fish larger, closer coastal concentrations and so on. As the high-quality resources are depleted, lower quality resources are used. This principle is well understood in economics based on work conducted 200 years ago by David Ricardo, and is called the principle of diminishing returns. High-quality sources require less effort to obtain what you want than low-quality resources, so depletion makes it harder and harder to obtain resources. The net result of this is depletion, i.e., humans have tended to use the higher grades of resources first, so that over time average grades decline. This concept goes way back to antiquity, where very high grades of copper in, for example, Crete, helped start the human exploitation of metals. On the other hand there is widespread belief that technological change can allow the exploitation of ever lower grades of resources. This issue will be examined more below and in the next chapter. But with respect to the Barnett and Morse analysis Cleveland (1991) showed that their analysis was seriously incomplete. The main reason that decreasing concentrations and qualities of resources was not translated into higher prices for constant quality final products was because of the increasing use of energy to exploit increasingly lower grade ore or other reserves in the USA and elsewhere. (Also there is some circularity in computing inflation rates in the original paper, too.) During the period of Barnett and Morse's analysis energy prices had been declining, so even though more energy was being used this was not reflected in higher prices. Thus, although economists have argued that natural resources are either not important or of decreasing importance to the economy, the truth is that it is only *because* of the abundant availability of fuels that economics can assign them low monetary value despite their critical importance to economic production. And the compensation of declining resource quality through increased exploitation rate of remaining resources almost always means that more energy will be used per unit of the resource, be it copper or oil, delivered to society.

The relative importance of depletion versus technological change for fuels can probably be best understood through the analysis over time of net energy and especially EROI. Net energy is the energy left over when exploiting a resource after the energy required to get that energy has been subtracted. EROI, a related concept, is simply the energy gained from an energy-obtaining activity compared to the energy used to get that energy (this idea is developed in more detail in Chap. 6 and the next chapter). If the EROI is declining it probably means that depletion is more important than technological improvements, and vice versa. While the concept of EROI is very simple, undertaking the actual analyses to estimate its value is often difficult for many reasons: because governments have been uneven in maintaining the required data base, because costs, including energy costs, are often closely held secrets by corporations, because the largest oil production is from national oil companies that tend not to make data available, because information is usually kept in monetary rather than energy terms, because it is necessary to correct the data used for its quality, i.e., the high-quality electricity generated from a PV system versus the fossil energy used to build it, and because of a lack of agreement on what boundaries are appropriate for the analysis. These can be large but surmountable problems. Also, EROI should not be confused with conversion efficiency, where energy already obtained is converted to some other form, such as coal to electricity.

10.3 What We Know About EROI Values and Trends for Different Fuels

Some countries maintain good enough energy use databases so that it is possible to undertake analysis, and if reasonable protocols are followed fairly good analytic results are possible and sensitivity analysis can be used to examine uncertainty. For example, Norway began finding major deposits of oil in the late 1960s in the North Sea, maintains good records, and has been well analyzed by Grandell et al. (2011; Fig. 10.2). It is a good example of what has happened around the world. Production started in 1970 and peaked in about 2000. Energy required to find, develop, and produce the oil continued to increase over the time of their analysis. The EROI of Norwegian oil peaked in 1996 at 45:1, that is, one barrel of oil (or 1 Joule—probably most of the invested energy was natural gas) invested in obtaining oil generated about 45 barrels (or Joules) in return at the well head (Fig. 10.2). By 2007 that number had declined to about 20:1. If the energy in the natural gas co-produced is included the numbers decline from 60:1 at the peak to 40:1 in 2007. The production and hence the EROI continued to decrease until about 2012 but then flattened out at a lower level, although financial and presumably energy investments have continued to increase substantially until at least 2014–2015 (Norwegian Petroleum 2016). Most studies of EROI show an increase with a new resource being developed followed by a peak and decline. A secondary pattern is that when prices and effort are high EROI tends to decline relative to the trend (e.g. 1970–1980 in fig. 10.3).

Similar patterns, although with lower EROI, have been observed for conventional oil and gas in the United States, the United Kingdom, Mexico, China, and for all privately traded companies (Gagnon et al. 2009; Guilford et al. 2011; Hall et al. 2014; Figs. 10.2, 10.3, 10.4 and 10.5). Hall and Cleveland (1981) found that in years when exploitation intensity is high (i.e., when high prices lead to more drilling) the EROI tends to decline relative to the time trend, and this pattern has continued. This means that, at least with conventional oil, it is not possible to increase production simply by drilling more wells. You have to drill thoughtfully rather than frequently in order to yield ample net energy. In time EROI appears to drive oil prices (King and Hall 2011); as EROI declines prices tend to rise, although other factors such as drilling rate may intervene. Brandt et al. (2015a, b) examined 40 large oil fields globally and found that the EROI varied enormously from 2 to 100 barrels returned per barrel invested. Many values were from 20 to 40:1, although since they did not include indirect energy costs comparable values with what we have reported might be closer to 15–30:1.

With respect to oil grades or quality of resources for oil, important issues are: onshore versus offshore, shallow versus deep, large fields versus small fields, light sweet versus, heavy sour, and so on. When new high quality oil fields or regions are found then they will be exploited. The problem is that this has been an increasingly rare experience over the past 50 or so years, and the average size of oil fields that we exploit keeps decreasing over time (Nehring 1981; Skrebowski 2016). These smaller, deeper, poorer grade fields tend to require more and more energy per unit

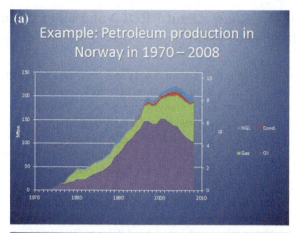
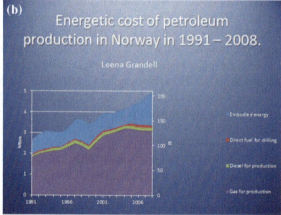
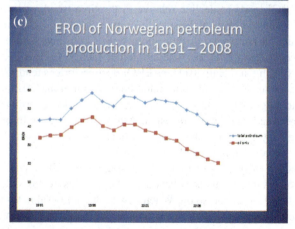

Fig. 10.2 **a** Production of petroleum from the North Sea fields in Norwegian waters, 1970–2008. The North Sea is the largest oil region found since the 1960s. Production has continued to decline through the middle of the next decade. **b** Energy cost for producing petroleum from the North Sea fields in Norwegian waters, 1991–2007. **c** EROI of Norwegian Petroleum production 1991–2008. Total production (*upper line*) includes the gain from natural gas (from Grandell et al. 2011)

10.3 What We Know About EROI and the Trends for Different Fuels

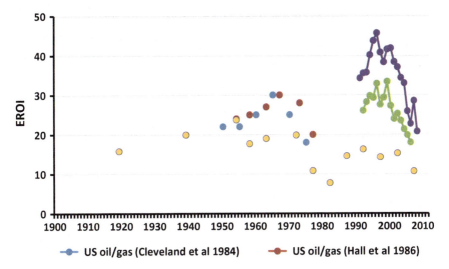

Fig. 10.3 A number of EROI studies for oil and gas. *Cream colored points* are for United States from Guilford et al. (2011); *Red* for US (Hall et al. 1986); *Blue* for U.S. from Cleveland (1995); *green* for all publically traded companies globally (Gagnon et al. 2009); *Purple* for Norway (oil only) from Grandell et al. (2011)

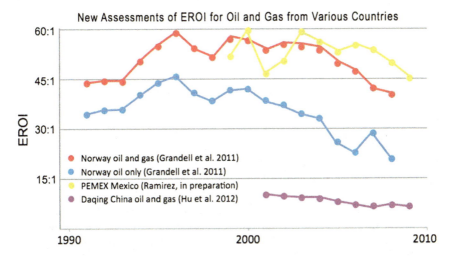

Fig. 10.4 A number of EROI studies for oil and gas. *Cream colored points* are for US from Guilford et al. (2011); *Red* for US (Hall et al. 1986); *Blue* for US from Cleveland (1995); *green* for all publically traded companies globally (Gagnon et al. 2009); *Purple* for Norway (oil only)

fuel delivered, decreasing the EROI over time. In general decreasing EROI is associated with higher prices (King and Hall 2011), as reflected in the generally increasing price of oil at the wellhead over time. We believe that as the EROI for

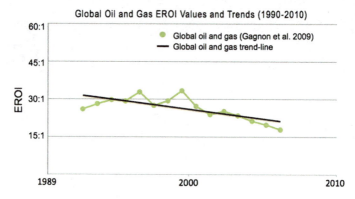

Fig. 10.5 EROI for global oil with possible projection

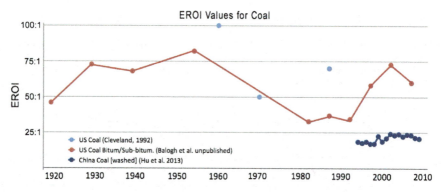

Fig. 10.6 EROI for US and Chinese coal production derived from Cleveland (1992), Balogh et al. (unpublished data) and Hu et al. (2013)

our principal fuels continues to decline the implications will be enormous as more and more of our total energy output, and hence our total economic activity, is diverted to get the same quantity of fuels (Fig. 10.1). The normal response by many people, including most economists, to this issue is that substitutes will occur and that technological processes will continue to improve so that there should be little or no concern. In fact there is a continuing race between technological progress and depletion. If the declining EROI of, for example, United States' oil is used as a yardstick it would appear that depletion is winning the race. While substitutes have in fact occurred to replace U.S. petroleum since its peak in 1970 the majority of this has been imported oil or, more recently, expensive "fracked" oil, having peaked and in moderate decline, and U.S. natural gas, which is subject to the same peaking and depletion issues.

An unresolved issue at this time is whether technical improvements in drilling efficiency, especially in fracked fields in, e.g., "light tight" formations in Texas and North Dakota, can decrease the energy and financial cost sufficiently to compensate

10.3 What We Know About EROI and the Trends for Different Fuels

for their original lower quality. Some assessments suggest that fracked oil has an EROI similar to that of US conventional oil (Waggoner 2013), others, e.g., Brandt et al. (2015a) suggest a somewhat higher, although decreasing value (a peak of 35:1 in 2010 with wide variance between wells, decreasing to 29:1 in 2013). Given that most fracking companies did not make a profit even when oil was $100 a barrel (Berman 2015) it does not seem that the EROI could be so very high, but Brandt's study seems well done.

A summary of existing studies that have been conducted (not always using the same procedures) show a very large difference in EROI for different fuels (e.g., Figs. 10.7 and 10.8). The fuels that are most commonly used tend to have high EROIs, and include coal (Fig. 10.6), hydroelectric power, and, to a lesser degree, oil and gas. A second trend seen in the data is that over time the EROIs of well-developed fuels tends to decline. This is most obvious for U.S. domestic oil, but is probably true for most others. Thus while it is true that there is a great deal of

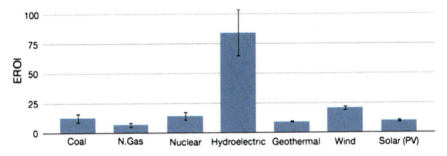

Fig. 10.7 Mean EROI (and standard error) values for known published assessments i.e. electric power generation systems. Values derived using known modern and historical published EROI and energy analysis assessments and values published by Dale (2010). See Lambert et al. (2014) and Hall et al. (2014) for further interpretation and detailed list of references

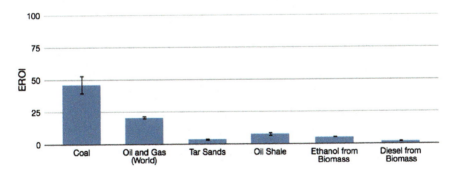

Fig. 10.8 Mean EROI (and standard error bars) values for thermal fuels based on known published values. Values are derived using known modern and historical published EROI and energy analysis assessments and values published by Dale (2010). See Lambert et al. (2013) for a detailed list of references. *Note* Please see text for discussion as all these values should not necessarily be taken at face value

low-grade oil left in the ground it is unlikely that the majority will provide a substantial net energy surplus.

EROI analysis is important for many reasons but perhaps especially because it puts a limit to the often stated economic principle that as oil (for example) gets scarcer the price will go up and then lower quality reserves will become economic —indefinitely. Instead at some point it will cost a barrel of oil or its equivalent to get a barrel of oil, and then no matter the price, those reserves will not be worth exploiting for energy. In the meantime as EROI declines there will be less and less energy available to run every other component of economies.

References

Balogh, S., M. Guilford, M. Arnold, Hall, C., 2012. EROI of US coal. unpublished data.
Barnett, H., and C. Morse. 1963. *Scarcity and growth: the economics of natural availability.* Baltimore, M.D.: Johns Hopkins Press.
Berman A. 2015. The Oil Price Collapse Is Because of Expensive Tight Oil. Posted in The Petroleum Truth Report on April 3, 2015. http://www.artberman.com/the-oil-price-collapse-is-because-of-expensive-tight-oil/.
Brandt, A.R., T. Yeskoo, and K. Vafi. 2015a. Net energy analysis of Bakken crude oil production using a well-level engineering-based model. *Energy* 93: 2191–2198.
Brandt, A.R., Y. Sun, S. Bharadwaj1, D. Livingston, E. Tan and D. Gordon. 2015b. Energy Return on Investment (EROI) for forty global oilfields using a detailed engineering-based model of oil production. *PLoS ONE* 10(12): e0144141.
Cleveland, C.J. 1991. Natural resource scarcity and economic growth revisited: economic and biophysical perspectives. In *Ecological Economics: The Science and Management of Sustainability.* 289–317.
Cleveland, C.,1992. Energy quality and energy surplus in the extraction of fossil fuels in the US. *Ecological Economics* 6, 139–162.
Cleveland, C.J. 2005. Net energy from oil and gas extraction in the United States, 1954–1997. *Energy* 30: 769–782.
Cleveland, C.J., R. Costanza, C.A.S. Hall, and R. Kaufmann. 1984. Energy and the United States economy: a biophysical perspective. *Science* 225: 890–897.
Dale, M., 2010. Global Energy Modeling: A Biophysical Approach (GEMBA). University of Canterbury, Christchurch, NewZealand.
Gagnon, Nate, and C.A.S. Hall. 2009. A preliminary study of energy return on energy invested for global oil and gas production. *Energies* 2: 490–503.
Grandall, L., C.A.S. Hall, and M. Hook. 2011. Energy return on investment for Norwegian oil and gas in 1991–2008: Sustainability: Special Issue on EROI. 2050–2070.
Guilford, M., C.A.S., Hall, P. O'Conner, and C.J., Cleveland. 2011. A new long term assessment of EROI for U.S. oil and gas: Sustainability: Special Issue on EROI. 1866–1887.
Hall, C.A.S., and C.J. Cleveland. 1981. Petroleum drilling and production in the United States: Yield per effort and net energy analysis. *Science* 211: 576–579.
Hall, C.A.S., C.J. Cleveland and R. Kaufmann. 1986. Energy and Resource Quality: The ecology of the economic process. Wiley Interscience, NY. 577 pp.
Hall, C.A.S., J.G. Lambert, S.B. Balogh. 2014. EROI of different fuels and the implications for society. Energy Policy. 64: 141–152.
Hu, Y., C.A.S. Hall, J. Wang, L. Feng, A. Poisson. 2013. Energy Return on Investment (EROI) of China's Conventional Fossil Fuels: Historical and Future Trends. Energy. 2013, 54:352–364.

References

King, C. and C.A.S. Hall. 2011. Relating financial and energy return on investment: Sustainability: Special Issue on EROI. 1810–1832.

Lambert, J., C.A.S. Hall and S. Balogh. 2013. EROI of Global Energy Resources: Status, Trends and Social Implications. Report to Division of Foreign Investment, United Kingdom. 136 pp.

Lambert, J., C.A.S. Hall, S. Balogh, A. Gupta, M. Arnold. 2014. Energy, EROI and quality of life. Energy Policy Volume 64:153–167.

Lovering, T. 1969. Mineral Resources from land. In National Academy of Sciences. Resources and Man, W. H. Freeman, San Francisco, 109–134.

Nehring, R. 1981. *The discovery of significant oil fields in the United States*. Santa Monica: Prepared for the US Central Intelligence Agency. Rand Corporation.

Norwegian Petroleum. 2016. Investments and operating costs. http://www.norskpetroleum.no/en/economy/investments-operating-costs/.

Skrebowski, C. 2016. Megaprojects: update as of 2016.

Waggoner, E. 2013. Sweet Spots, EROI and the limits to Bakken Production. MS Thesis, State University of New York College of Environmental Science and Forestry.

Chapter 11
Methods and Critiques for EROI Applied to Modern Fuels

This chapter gives the basic methodology for calculating EROI, discusses some controversial or unresolved issues and confronts a number of critiques that have been laid against the technique, both generally and with respect to particular applications. EROI analysis is not precision science, but with attention to the problems and uncertainties I believe that reasonably accurate and repeatable results, with uncertainty included, can be obtained. I hope by dealing with these issues explicitly the reader will be reassured that it is quite possible, although sometimes difficult, to derive meaningful and useful EROI analyses, and that this could be undertaken much better with a relatively modest governmental commitment to collecting the needed data. What we cannot do, however, at least yet, is to answer definitively certain philosophical issues pertaining to the boundaries of the analysis.

11.1 How We Do EROI Analysis: More Detail

For those seeking to actually undertake EROI analysis I recommend that you should start with Table 11.1 and then consult the Murphy et al. (2011) protocol paper. These give a general overview as to how to proceed. One particular advantage of the Murphy et al. protocol is that it recommends a standard approach (energy at the wellhead, farm gate, bussbar or other source), called "standard EROI" or $EROI_{st}$—which allows for a comparison among different fuels/sources—as well as a procedure for undertaking other approaches that the investigator might favor. That paper also has a long section on correcting for energy quality, i.e. the difference between coal or gas and electricity.

Table 11.1 Step by step instructions for EROI analysis

The objective of this table is to give a short, unambiguous procedure for conducting a basic EROI analysis. Please see also Murphy et al. 2011 (Modified from Hall 2015)

Step 1. State objectives

The first step in performing an EROI (or any) analysis is to state the objectives of the analysis clearly. This will allow the reader to get a sense of the scale of analysis being performed and whether or not there are other analyses with similar objectives.

Step 2. Create a flow diagram and identify system boundaries

Figure 11.1 represents a generic flow diagram for any energy production system, and can be used as a reference. The symbolism developed by Odum (1983), Hall and Day (1977) for systems flow diagrams, although not used here, is recommended when drawing the flow diagram for any complex EROI analysis (see e.g. Fig. 2.5 and Fig. 1 in Murphy et al. 2011). All direct, indirect, and embodied energy inputs and outputs should be included in this flow diagram, as well as the boundaries used for analysis or sensitivity analysis. Prieto and Hall (2013) can be examined as a guide to the larger boundaries that might be considered.

Step 3. Identify all appropriate inputs and outputs within system boundaries

Once the flow diagram is complete, the various flows of energy, defined by arrows connecting two symbols, should be identified, estimated where possible, and labeled on the flow diagram as either direct, indirect, or embodied energy inputs or outputs. I recommend using the concepts and nomenclature developed in Murphy et al. 2011.

Step 4. Identify and convert financial flows if necessary

When direct flow measurements are not available, often times only financial data is available for calculating energy flows, such as the cost of transporting oil or developing and installing machinery. This data needs to be converted to energy units using an energy intensity value (e.g. as outlined in Murphy et al. 2011 which can be referenced for a fuller discussion of how to derive energy intensity values and which intensities are appropriate for various analyses. Once these financial flows are identified, convert them to energy units using whichever energy intensity value was chosen.

Step 5. Make the calculation

Once steps one through three are completed, the analyst should have sufficient knowledge to identify which specific EROI calculation is being performed as, or in addition to, $EROI_{st}$. At this point the analyst should identify the specific EROI calculations that they are performing and define the EROI equation by placing the appropriate energy flows from the flow diagram developed in step 1 into the numerator and denominator of the EROI calculation.

Step 6. Calculate EROI

Solve the simple equation to derive the EROI ratio. Provide straightforward assessment of the limits to data as well as sensitivity analysis of uncertainties. Ideally come at the assessment using several data sources or boundaries to analysis and examine how much difference this does or does not make.

Step 7. Choose method of energy quality adjustment as part of sensitivity analysis

All of the energy inputs and outputs should be undertaken with both heat equivalents and quality-adjusted energy if possible. We recommend using a price-based aggregation or a Divisia approach for quality adjustments (see text), unless there is a good reason for doing otherwise (see Cleveland et al. 2000; Murphy et al. 2011). At a minimum, electricity should be multiplied by a

(continued)

11.1 How We Do EROI Analysis: More Detail

Table 11.1 (continued)

factor of 2.6 or 3 to represent mean thermal requirements. The analyst should spend time identifying the benefits and shortcomings associated with whichever method is chosen, including any underlying assumptions. For example, if an EROI analysis uses mainly one energy type for inputs and outputs, then a quality adjustment adds little value.
This is the final step in the process where all of the energy units are aggregated and the EROI value is calculated and sensitivity analysis for energy quality or data uncertainties are undertaken. Each EROI analysis should have at a minimum the $EROI_{stnd}$ as well as any other EROI calculations of interest. The investigator can then compare his or her $EROI_{snd}$ with others, and indicate whether and how an alternate EROI assessment adds useful information.

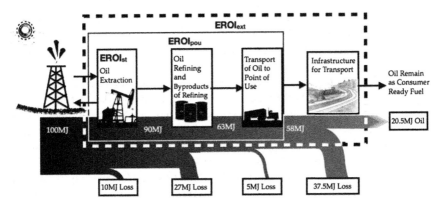

Fig. 11.1 Boundaries of various types of EROI analyses and energy loss associated with the processing of oil as it is transformed from "oil at the well-head" to consumer ready fuels (figure from Lambert and Lambert (in preparation) based on calculations by Hall et al. (2009))

Mathematically EROI is simply the ratio of an output divided by an input. Normally it should be applied only to energy extracting from nature, in order to evaluate net energy delivered to society, i.e. the gross energy output of the process minus the energy used to get that energy, usually measured at the edge of the facility:

$$EROI = \frac{\text{Energy returned to society from an energy gathering activity}}{\text{Energy required to get that energy}}$$

Normally EROI analyses do not include the conversions after the energy is obtained. The "energy required" in the denominator represents energy withdrawn or diverted from society, and includes the energy used by businesses to make drilling or digging equipment, or extracted fuels that are used on site rather than delivered to society. Simple enough, but how does one get data to undertake such an analysis?

11.2 Energy Return Data

Usually energy output data (for a well, a field or a nation) is published in regular governmental or industry reports, sometimes vetted, and is presumably reasonably accurate. This information for all major fuels is available from various national energy statistics agencies, such as the U.S. Energy Information Agency (EIA). National data published by the EIA is usually considered fairly accurate, and it tends to agree (to within a few percent) with other data such as that published by the International Energy Association (IEA) or by BP (formerly British Petroleum). A great deal of output data for other countries is maintained by Jean Laherrere, a very knowledgeable and well-connected oil professional, once head of exploration for Total, France's principle petroleum company. He publishes this at national, regional and individual well data, on the France-ASPO web site. The EIA also publishes such data by State and by week as well as data on natural gas and coal: (http://www.eia.gov/dnav/pet/pet_sum_sndw_dcus_nus_4.htm).

One particular problem with output data, however, is that definitions may change over time. For example the EIA data on oil increasingly includes not just what is normally called "conventional" petroleum but also biomass liquids, "refinery gains" and of greatest importance "natural gas liquids", which are low molecular weight liquids (pentanes, hexanes etc.) produced along with natural gas. Since, unlike conventional oil, natural gas production is still increasing, the production of natural gas liquids is also increasing. These liquids, which today make up nearly 10 % of the category "all liquids", are very useful as petrochemical feedstocks but much less so for e.g. transportation. Their increase over time gives a false sense that production of genuine oil is still increasing when, essentially, it is not. Thus at a minimum the output of these various things need to be converted to similar energy units. Another issue is that it is rarely stated explicitly whether the data is gross or net (i.e. whether the fuel used on site has been subtracted from the output). Based on definitions given this seems to be the case for e.g. U.S. gas (called "marketable gas"), but not clearly so for oil. Sometimes data is available for individual wells and projects. The State of North Dakota, for example, maintains very good records on individual oil and gas wells.

11.3 Energy Investment Data

Deriving energy input (investment) data for a project, field or nation is more difficult and less certain. In general one should attempt to get energy data directly. However it is sometimes difficult to get the desired cost of energy itself for estimating EROI because they are often closely held secrets by corporations and national oil companies that tend not to make much data of any kind available, and because relatively few nations maintain industry-by-industry energy cost data.

There are two main types of energy investments: direct energy, which is the energy used on site, such as natural gas to run drills or pumps or provide pressure down-hole, and indirect energy, which is energy used off-site to derive materials or services used on site or in other places. The latter can be quite complex, for example the steel used for a drilling rig or the manufacture of a drill bit is pretty straight forward, but how about including an energy cost of business services, or land purchases or for taxes? Of even more ambiguity, should one provide for the goods and services purchased by a worker's paycheck?

There are three principal procedures used based on sources of data (we use oil and gas extraction as an example):

11.3.1 National Energy Accounts of Direct Energy Used

Canada, China, England, Mexico, Norway and the United States maintain records of the energy used by various industries including energy industries. It seems to me that professionally derived information at the national level is the best source of information, but there are no studies to confirm or deny that. For example, the United States Government Department of Commerce maintains and publishes data in the *Census of Mineral Industries* at five year intervals from 1954 to 1997. Other data is available back to 1919 and on line since 1997. For example, Standard Industrial Code sector 13, "Oil and gas extraction" includes data on energy used by "all firms that explore for oil and gas, drill oil and gas wells, operate and maintain oil field properties that produce oil and gas, and all other activities in the preparation of oil and gas up to the point of shipment from the producing property". Sector 13 also includes firms engaged in producing natural gas liquids derived from oil and gas field gases. These data include the direct (i.e. on site) energy cost for extracting petroleum, including the fuel and electricity used in oil and gas fields to run pumps, compressors, trucks and so on, the coal, crude oil, natural gas, and refined liquid fuels such as gasoline, residual, and distillate fuel used, as well as purchased electricity (and electricity generated by captive fuel use which is excluded because including it would double count the fuels used to generate it). It is derived by government experts undertaking surveys with individual industries. It used to include uncertainties in the estimates, which typically have been about 5 % of the mean. This data has been used by e.g. Hall and Cleveland (1981), Cleveland et al. (1984), Hall et al. (1986), Cleveland (2005) and Guilford et al. (2011, see especially the Appendix). Grandell et al. (2011), Hu et al. (2013) used somewhat similar data for Norway and China.

It takes some serious time investment, a good (university or governmental) library and the internet, and ideally some help from the professional staffs to do such analyses. In the U.S., the completeness and possibly the accuracy of the published data appears to be degrading over time, as the number of people deriving the data has been cut along with Federal budgets. Nevertheless we have found the staff of the U.S. Bureau of Census to be very helpful.

Sometimes there are assessments of energy used for a particular facility, such as an oil well or a PV facility, where the output is measured directly.

11.3.2 National-Level Accounts for Capital Expenditures and Other Indirect Uses

An estimate of ALL the energy used to get that oil and gas becomes more complex. The indirect energy used includes energy used off site to produce and maintain the capital equipment used to extract e.g. petroleum, coal or other energies, such as to make steel forms, drill bits, wind turbines and blades, photovoltaic panels and so on. The indirect energy cost per ton of common materials such as concrete or steel is usually known (see next). Other forms of capital are more commonly calculated from data on the dollar cost of purchases of those inputs to the extraction industries from e.g. the *Census of Mineral Industries*. Materials include the purchase of chemicals, wood products, steel mill shapes and forms, and other supplies "used up" each year in e.g. the coal and petroleum industries, or the dollar value of capital depreciation in SIC (Standard Industrial Code) sector 13 (oil and gas extraction) (see Guilford et al. 2011, especially Appendix).

The indirect energy cost of these capital and materials then can be calculated as the dollar cost of capital and materials times the *energy intensity* of the formation of capital and materials (Joules/$). The energy intensity of capital and materials is measured by the quantity of energy used to produce a dollar's worth of output in the industrial sector of the economy. The mean value of that ratio for all economic production in a country is the ratio of fossil fuel and electricity use to real GNP produced by industry e.g. for the U.S. in 2005 8.3 MJ/$, the mean for the society as a whole (this needs to be calculated each year). A better value is an estimate for industrial production of 14 MJ/$ per dollar with a high value of 20 MJ/$ for the oil and gas industry (see Murphy et al. 2011). More explicit analysis can be done using energy input-output analysis (see below) where this is possible. In the U.S. after 1972 the energy associated with producing and supplying these indirect costs often were as high or higher than the direct use.

Using dollar values is not an ideal measure of capital input because it reflects financial variables (i.e. inflation and deflation in prices) in addition to actual physical costs. However, investments and capital depreciation are the only aggregate measure of capital input for the petroleum industry, and despite their shortcomings it serves as an approximate indicator of the trend in capital use over time. As with energy outputs it requires some serious time and resources to undertake such analyses. Mulder and Hagens (2008) discuss how increasing the comprehensiveness of an assessment usually means decreasing precision. A government guide to undertaking such analyses would be very helpful. A critical issue with assessing indirect costs is the boundaries used. This is considered below.

11.3.3 Process Analysis

A third approach, or set of approaches, to determining EROI, or something like it, is called *process analysis*, or sometimes (mostly in Europe) *life cycle analysis* (LCA). Both of these terms refer to conceptual approaches originally designed for environmental issues, although that may be specifically tailored or modified to generate EROI analysis. Process analysis refers generally to a general set of instructions for undertaking a task, from building a sand castle to writing a business plan to creating a power plant. Depending on the data available, energy requirements may be calculated per unit mass or per dollar value of the input. Life cycle analysis (LCA) of energy is a related area of research, which seeks to quantify the resource use and/or environmental impacts associated with energy supply or use, or more generally, environmental impact associated with the various stages of manufacturing and using a product, including extraction and processing materials, manufacturing, use and disposal. This approach has been modified to focus on energy used at each step (e.g. Carbajalis Dale et al. 2013; Hertwich et al. 2014). In recent years authors have frequently used cumulative energy demand (CED) from LCAs or computer programs and data bases such as Ecoinvent to determine (or define) EROI or something like it (e.g., Kubiszewski et al. 2010; Fthenakis and Kim 2011; Carbajalis Dale and Benson 2013). See Murphy and Carbajalis Dale (2016) for an attempt to apply the best features of LCA to EROI analysis. In practice Arveson and Hertwich (2015) argue against placing too much faith in the comprehensiveness of LCA for EROI studies. I interpret that to mean that many LCA analyses do not include all of the actual energy inputs required.

In theory, process analysis provides the most detailed information on the energy cost of goods and services. There are several practical problems, however, such as inadequate data, and incomplete boundaries that effectively limit its applicability. In addition there are not energy costs calculated for many of the real inputs.

Both process analysis and LCA, as well as other methods for indirect costs, are based on getting a relatively complete "shopping list" of goods and services required to undertake a particular project, say constructing an oil well or PV module, and the existence of the energy requirements to generate each item. For example, Hall et al. (1979) got a very detailed listing of all the inputs required to construct a proposed coal generating plant in central New York State (and also a regional insulation program). Each item on the list was given in physical units, such as tons of concrete or steel, hours of bulldozer activity and so on, or as dollar costs. Then energy assessments were made on the energy cost of each item and all were summed to get the energy cost of building the facility. This approach assumes that you know or can get energy costs for each item. When Hall et al. undertook their analysis, there had recently been a comprehensive energy I-O analysis undertaken at the University of Illinois (e.g. Herendeen and Bullard (1975), Hannon (1981)), which calculated the energy cost of a dollar's worth of product from each sector of the U.S. economy, including the energy cost of the chains of inputs to each manufacturing sector. Thus they could estimate the energy cost of all components, from site preparation to turbines to electrical gauges from their dollar costs, which they had.

The University of Illinois input-output table breaks the economy into about 400 different sectors and required a great deal of industry-specific data that needed to be derived from Census of Manufacturers' data. The numbers derived represent the quantity of direct plus indirect energy that each industry purchases from all other sectors in order to manufacture a dollar of final demand product. The energy intensity factor (Btu/$) of each good or service can be calculated from the dollar flows between industries. These results give a comprehensive and reasonably accurate representation of both the direct and indirect energies used to manufacture a product and is presumably more comprehensive than process analysis. It is much more difficult to undertake such studies now as the requisite energy I-O analyses are not routinely or comprehensively undertaken. Some estimates for about 2000 are available from Carnegie Mellon's green building program as summarized in Murphy et al. (2011).

In our opinion even such attempts to generate comprehensive energy costs miss important costs. Prieto and Hall (2013) attempted to circumvent this problem by following all of (as best they could) the money spent for several large solar PV sites in Spain, something that was possible because of Prieto's experience of being chief construction engineer on a large PV project, who had to sign for every Euro spent for each and every input and output. They then assigned an energy intensity to each category. The mean energy use for the Spanish economy in 2008, the year for which they did the analysis, was 7.16 MJ per Euro, and Prieto and Hall, based on Murphy et al. (2011), assumed that the ratio was twice that for manufactured goods or engineering, and one third that for business services. Interestingly, they derived about the same energy cost per GW when they took all money spent times the national mean (7.16 MJ/Euro) as they found when they did a very detailed analysis item by item. This is consistent with the view of Hannon and Bullard that when one purchases a complex product from final demand all the different energy intensities tend to "come out in the wash". A major issue remaining is whether other energy costs, such as the energy required to support a laborer's paycheck, should be included. This is considered below.

11.4 Quality Corrections

While all fuels can be measured by the amount of heat they generate when completely combusted or otherwise used, different types of energy have different qualities that are valued differently for different purposes. Oil is usually a superior transport fuel than gas or coal, electricity can do many things per heat unit that fossil fuels cannot and so on. In general humans are willing to burn oil, gas and coal to generate a lesser quantity of more highly valued electricity, "throwing away" half to two thirds of the heat content in the process. Thus we might say that one heat unit

11.4 Quality Corrections

of electricity is worth about three units of fossil fuel, since society is willing to trade three heat units of fossil fuel for one of electricity.

A relatively easy way to account for this difference in quality is to assume that price differentials per heat unit captures the value of different fuels. This approach has been formalized in the *divisia index*, which can be examined further in (Cleveland 2010; Murphy et al. 2011; Department of Energy 2000). For a large complex projects one can weigh and sum inputs and outputs accordingly:

$$EROI = \sum_{i=1}^{n} E_{i,t}^o / \sum_{i=1}^{n} E_{i,t}^c$$

where E^o and E^c are the energy output and input, respectively, of energy type n at time t, measured in thermal equivalents. The qualities of different fuel types, such as primary electricity or oil vs coal can be corrected for by weighting fuels by price (i.e. E_i times P_i) or some other indication of relative quality.

11.5 Estimating EROI at Point of Use

The original EROI calculations, and the majority since, have been for EROI at the source of the energy, be it well head (for oil and gas), buss bar (for electricity) or farm gate (for e.g. food or biofuels). In some respects this makes the most sense because we are interested in the cost of getting energy at its source. But a reasonable question is what would be the EROI of a fuel at the point where it is used, since there may be very different efficiencies for different fuels between the source and the point at which it is used. Unfortunately such studies are rare, and we must remember to always start with a source of energy from nature. Otherwise we are just doing internal efficiencies, not EROI.

One such study that did undertake a fairly thorough analysis was undertaken for oil by Hall et al. (2008). They were interested in the energy cost of getting one unit of gasoline into a car or truck and then using it. They undertook their analysis by starting with 100 MJ (about 3 L) of oil coming out of the ground. At an EROI of 10:1 (roughly what might be found in the US about now) 10 MJ were used (or diverted from society) to get that 100 MJ (or 90 net) out of the ground. Then another 10 or so was used to refine it, and there was another loss of 17 MJ as various petrochemical products that could not be used for fuel (for example heavy fractions used for roadways), and a use of another 5 MJ to transport the fuel to its point of use. Thus of the 100 MJ started with only 58 {100 minus (10 + 10 + 17 + 5)} made it as fuel to the point of use. Their study also considered the cost of maintaining the vehicles, bridges and roadways (i.e. the depreciation cost) pro rated for the use of the gasoline. In energy terms this cost another 37.5 MJ, leaving only 20.5 MJ to put in the car or truck. Thus (giving credit for the other coproducts derived from the original oil) the authors concluded that the energy return on investment at the well head would have to be at least three to one if that oil were to be used to drive a car or truck. This implies that fuels with very low EROI, such as corn-based ethanol (with an EROI estimated as at most about 1:5:1, and more probably very close to 1:1) could not be used to operate a transportation system, let alone a modern civilization.

11.6 EROI of Obtaining Energy Through Trade

Kaufmann (Chap. 8 in Hall et al. 1986) derived a procedure that can be used to estimate the EROI for U.S. imported oil from the perspective of the importer. Kaufmann found that before the oil price increases of the 1970s, the EROI for imported oil was about 25:1, very favorable for the United States, but that dropped to about 9:1 after the first oil price hike in 1973 and then down to about 3:1 following the second oil price hike in 1979. The ratio returned to a more favorable level (from the perspective of the U.S.) since then because the price of exported goods has increased through inflation more rapidly than the price of oil. Subsequently the ratio has (presumably) increased and decreased in response to the relative prices of oil compared to exported goods and services.

Lambert et al. (2013) modified the concept slightly to use it for developing countries. This is particularly important for such countries as Costa Rica and Pakistan that are highly dependent upon oil but have no domestic resources. Since oil can be purchased only with dollars, not local currencies, an oil importing nation must generate goods and services to export to gain foreign exchange (i.e. dollars). But energy must be used in that country to make those goods and services. In other words one pays for the chemical energy in imported oil with the embodied energy of exports (Fig. 11.2).

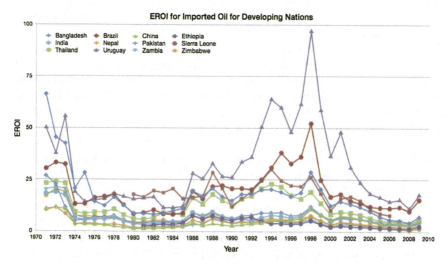

Fig. 11.2 EROI for imported oil for developing nations. See text for derivation. (From Lambert et. al 2013)

They found, for example, that in 2009 Pakistan imported 126.4 million barrels of oil which cost the country (i.e. its various buyers) some 7.5 billion US Dollars. To get these dollars, Pakistan exported 20.1 billion dollars' worth of various goods, such as textiles and clothing. That year, the economy of Pakistan as a nation used on average about 24 MJ of energy from all sources per dollar of economic production. For example, Pracha and Volk (2011) found that the energy intensity of Pakistani agriculture was similar to that for the United States. They assumed that the exports, which covered a range of products, required an average amount of energy per dollar to produce. Thus to get the 7.5 billion dollars needed to buy its oil, Pakistan had to use some 180 million GJ of energy (7.5 billion dollars times 24 MJ/dollar) to run its farms, factories and so on, an amount equal to some 30 million barrels of oil. Thus it took about one barrel of oil, used (or you might say invested) in Pakistan's economy, to generate enough dollars from external sales to buy 4 barrels of oil. Hence the EROI for imported oil for Pakistan in 2009 was about 4:1. This was a relatively unfavorable time for Pakistan, as earlier the ratio had been about 24:1, and it has also become somewhat more favorable since. The U.S. too, imports a great deal of oil (about half of its total consumption) and must pay for that in the same fashion. When the ratio is very unfavorable, however, many countries, including the U.S., tend to pay for the oil with debt, which is considered in the next chapter.

11.7 Methodological Issues, Problems and Criticisms of EROI

EROI seeks to compare the amount of energy delivered to society by a given technology or procedure to the total energy required to find and extract, and in some cases process, deliver, and otherwise upgrade that energy to a socially useful form, that is, the ratio of energy delivered compared to energy costs. This would seem to be a rather simple thing to do, at least in theory, but in practice its implementation is difficult and sometimes daunting. The range in EROIs published by different authors for nominally the same resource has led some to question the validity of the concept (e.g. Heinberg and Fridley 2016; Nate Hagens personal communication). I believe that most of the issues are because of different assumptions and procedures rather than some fundamental inconsistency with EROI, but I shall attempt to address each issue. Certainly important issues remain. I warn the reader that it is important to distinguish between criticisms that are well thought out and valid, usually by professionally trained and peer reviewed authors, and others that are irresponsible, poorly or inaccurately argued and/or are motivated by political or corporate interests (For example: see http://bountifulenergy.blogspot.com/2010/09/eroi-doesnt-matter.html).

Next follows three critiques in which these methodological issues are considered: one on the definitions used and two on specific energy systems (corn-based ethanol and photovoltaic) where different investigators have come up with different conclusions, and I examine the reasons for the differences. Finally I consider some other methodological issues not previously considered.

11.7.1 Criticisms of EROI Studies: Definitions

Engineer Carey King has presented several papers (e.g. King et al. 2015) in which he argues that in many EROI studies the terms are not precisely enough defined. He introduces a number of more explicit terms for what has been called more generally in the literature EROI. Specifically he believes it important to distinguish more carefully between energy returned on investment (which he calls Net External Energy Ratio or NEER) and power returned on investment (which he calls Net External Power Ratio). King suggests that if performing a calculation of EROI that has original units of (energy/yr output)/(energy/yr invested), then that is a ratio of units of power, and not energy.

My position is that given that EROI is a dimensionless ratio, the units in numerator and denominator are the same, and cancel whichever is used. King (personal communication) agrees that performing calculations using annual energy inputs and outputs for each year is perfectly legitimate for annual analyses, such as commonly used for petroleum (e.g. Cleveland et al. 1984; Guilford et al. 2011) (although energy per year technically is a power term). One has to be careful with rapidly growing or shrinking technologies (e.g. Neumeyer and Goldston 2016). King agrees that an assessment adding or integrating gains and costs over the lifetime of a project, as implied by the EROI derived by Prieto and Hall (2013), is properly EROI.

King also wants users to be careful about whether the energy output includes that invested in getting that fuel or not. "The difference between GEER and NEER, as well as between GER and NER, is that the denominator of the equation is subtracted from E_{ext} in the numerator of NEER and NER, but not GEER and GER." (Gross and net energy return: Effectively, NEER is mathematically equivalent to EROI as defined in this chapter). This is a good and legitimate question, and has not been handled sufficiently explicitly in the literature, including my own studies. For example considerable amounts of natural gas are used to pressurize and pump oil fields. Is this gas subtracted from the output (production) values used in analysis? Apparently, but not explicitly so. EROI values are normally presented as if the numerator is net output (e.g. Guilford et al. 2011). An examination of the definitions given with the data by the U.S. Energy Information Agency (2016) finds that despite considerable efforts at clarity and definitions it is not clear whether the basic production information they give is for gross or net (i.e. gross minus on site use) although it appears to be gross for oil but net for gas. It appears for gas as the difference (about 20 %) between the categories "gross withdrawals" and "marketed production" which is fuel "used on leases", and since most fuel used for production of oil and gas is gas it should be possible to undertake an estimate of net production.

11.7 Methodological Issues, Problems and Criticisms of EROI

In the meantime, since gas is the primary fuel used on site the numerator of most EROI equations should be considered net withdrawals.

Arveson and Hertwich (2011) also are concerned about definitions: "The core idea of EROI is to measure the relation of energy diverted from society to make energy available to society. CED (cumulative energy demand–used in life cycle analyses), on the other hand, includes forms of energy that are not appropriated by society, such as fugitive methane emissions from oil wells as well as losses of heating value of coal during transport and storage. Such energy forms should be excluded from EROI". They continue with some caveats about using life cycle analysis for determining EROI.

In general the way to deal with different perspectives is, as with many scientific assessments, *sensitivity analysis*, that is report the results under differing assumptions about data or philosophy and let the reader choose. Usually such analyses do not make dramatic differences with reasonable data. For example Guilford et al. (2011) found that when extreme assumptions were made about the energy intensity of indirect costs it changed their calculated EROI by about 25 %, and in no way changed the basic results of the study.

There have been two main areas where the differences in EROIs calculated between different groups of investigators have been especially large: the EROI of corn ethanol and the EROI of solar photovoltaic systems. I will examine the methodological issues from the perspective of these two issues, each of which has been characterized by rather rancorous exchanges.

11.7.2 Differences in Results: Corn-Based Ethanol

In the United States presently there is a 6 to perhaps 30 billion dollar per year Federal subsidy paid to U.S. farmers and processors to make "gasohol", mandated as a 10 % blend of alcohol with gasoline at most gas stations. There is considerable promotion of gasohol as a "domestic" fuel source that will free us from our dependence upon foreign oil (we currently import about half of our oil). On the other hand a number of investigators, most explicitly David Pimentel and Tad Patzek (e.g. Pimentel 2003; Patzek 2004, 2007; see also Murphy et al. 2011), have calculated that alcohol fuel from corn is an energy sink, meaning that it takes more energy to produce the fuel than is gained from its combustion. If it takes a gallon of gasoline to create less than that energy equivalent of ethanol then obviously it makes little sense to do so except perhaps to the farmer/processor who reaps the market cost plus the subsidy. On the other side of this argument are several investigators who have argued that if the "proper" methods of analysis are used then the EROI is at least 1.4 to one (see Farrell et al. 2006). There were extensive and sometimes acrimonious arguments among those concerned with this issue as to whether one side or the other is using the "proper" methods of analysis. Most generally the arguments revolve around the issue as to whether the fuels are a net energy source or not, the policy implication being that if the they are a net source the subsidy should continue, and if not then it should be discontinued. But in fact the subsidies, while decreased in 2011, continue on a large scale independent of the scientific arguments, and we are all forced to use gasoline that has been "diluted" with ethanol.

There have been relatively few studies that have attempted to understand the reasons for the different estimates of EROI for the same fuel. But one explicit study has been undertaken by Hall, Dale and Pimentel (2011). In about 2008 the latter two authors of this trio, each a competent and respected agricultural scientist, were well known for their acrimonious debate about whether biofuels delivered significant net energy or not. In earlier publications Dale and a coauthor had given an EROI of 1.73:1 for corn-based ethanol in the United States and Pimentel and a coauthor 0.82:1. In other words, one author had reported a moderately positive EROI and the other a negative EROI for the same fuel. Hall (myself) got these two authors to join me in the above-mentioned publication to try to examine what the reasons were for the differences. After considerable analysis we found the main reason, explaining about half the difference, was that Dale had attributed only about three quarters of the energy inputs required to the energy cost of producing the ethanol. One quarter was not counted to ethanol but rather to the residual mash, used as animal feed, that was "co-produced": in other words he did not attribute all the input energy cost to the ethanol because a valuable coproduct was also generated. Of the remaining difference about a quarter was explained by different estimates for e.g. energy cost per ton of fertilizers and a more inclusive list of inputs by Pimentel, i.e. a larger boundary. When similar assumptions and input data were used the results of the two earlier studies generated similar results. Which approach is "proper"? Pimentel agreed that for some purposes co products should be included, and all authors agree that it is important to include all required costs. In general it does not seem that there is explicitly only one way to do the analysis. If we look only at the fuel corn ethanol does not seem to be an energy source. If we look at the process of making fuel and animal feed it is a modest energy source. Take your pick!

11.7.3 Differences in Results: Photovoltaic Systems

There are very good reasons that society should leave or at least decrease its use of fossil fuels, either because of their depletion or because of concerns about CO_2 and climate. In the first decades of the twenty first century a number of studies gave EROIs of 6–10:1 (or more often expressed as energy pay back times of one or two years) for photovoltaic (PV) systems (e.g. Fthenakis et al. 2011; Raugei et al. 2012). These numbers were used by solar advocates to argue for the importance and economic viability of solar PV systems, and in some cases that solar PV systems were comparable to fossil fueled systems. But in 2013 Prieto and Hall came out with a much lower estimate of EROI of 2.45:1 for sunny Spain with sophisticated engineers, which caused a great stir amongst solar advocates. This was initially greeted with disbelief by many in the industry. But then similar results were

11.7 Methodological Issues, Problems and Criticisms of EROI

published by Palmer (2013) for rooftop PVs with battery back up in Australia, and Weissbach et al. (2013), for Germany (see also Raugei 2013; Weissbach et al. 2014; Raugei et al. 2015). In 2016 Ferroni and Hopkirk published an estimate of a below-unity EROI for cloudy Switzerland and Germany. Leccisi et al. (2016) and Raugei and Leccisi (2016) came back with estimates of values of 9:1 or higher. How could two different groups of competent investigators get such different estimates?

The most comprehensive assessment of all of the energy costs of providing energy is found in the book "Spain's Photovoltaic Revolution—The Energy Return on Investment" (2013) by Pedro Prieto and Charles Hall. The book differs from many earlier analyses in it attempts to include (nearly) ALL energy costs actually used, not just the costs of the modules and some related hardware, and it uses measured rather than estimated energy output. It is also different in that it is applied to the PV-technology in a country having a much stronger insolation than Switzerland, Germany or the Netherlands. Of particular importance is that Prieto and Hall attempted to calculate the complete energy used to support the PV system by "following the money", i.e. by attempting to assess all the money flows necessary for the system to operate (understood by Prieto because of his extensive on-site experience as Project Director, Project Designer, Consultant and Director of Development of a solar PV company). They then assigning an energy cost to each monetary cost using specific energy intensities: the mean energy use for the Spanish economy (7.16 MJ per Euro), and twice that for manufactured or engineering items, and one third that for business services as given in the protocol paper by Murphy et al. (2011). As we indicated above, they derived about the same energy cost when they took all money spent times the national mean (7.16 MJ/Euro, similar to the global mean) as they found when they did a very detailed analysis of 24 categories of items, including such things as energy costs of roads and cleaning, surveillance, business services, meetings attended by engineers as well as modules. This is consistent with the view of Herendeen and Bullard (1975) that when one purchases a complex product from final demand all the different energy intensities tend to "come out in the wash". Raugei and Leccisi (2016), for example, did not calculate any energy cost for which they could not get a direct energy measurement for in their assessment of PV and fossil fuel derived energy for England. To me this seems to miss some costs.

11.7.4 Corrections for Energy Quality

The largest difference in EROI between the two sets of investigators is based (depending on the publication) on corrections for quality between fossil fuels and electricity. Raugei et al. (2012) were very critical of comparing the apples of fossil fuels (where EROIs at the source were generally higher) with the oranges of higher quality electricity. They said that a number of summaries (e.g. the "widely cited 'balloon graphs' (Hall et al. 2008; Murphy and Hall 2010) and bar charts (Hall and Day 2009) have compared many technologies simply in 'heat equivalents', i.e. the energy values are given in terms of their abilities to heat water with no correction for energy quality". The fundamental issue is that since we are willing in society to trade about 3 heat units of coal, oil or gas to generate one heat unit of electricity, the EROIs of the electricity derived from PVs or wind turbines (or nuclear power plants) should be weighted by a value of some three times that of a heat unit of fossil fuels.

In some other publications these differences have been examined by looking at the energy efficiency of delivering e.g. electricity to the final user. This fossil fuel-derived electricity would be converted to electricity and the distribution losses would be figured in. For example Raugei and Leccisi (2016) examined the EROI of electricity systems for the United Kingdom, where the EROI for much-depleted coal at the mine mouth is much lower than for e.g. the United States, and finds the EROI of PV and wind systems somewhat higher.

In my view this is a legitimate criticism although we can quibble with parts of it. First, we certainly understand the concept and have in fact generally given such solar PV values in both thermal values and quality corrected values (for example: Cleveland et al. 1983, Table 1); Hall et al. 1986, p. 48); Hall et al. 2013 Fig. 3; Prieto and Hall 2014 (p. 116) where a quality factor of three was used for electric output, assuming all inputs were fossil fuels. Second, this can imply that all inputs to manufacturing PV systems are fossil fuels (although Raugei and colleagues correctly use the source energies for electricity) and that all electricity outputted are used for high quality functions. In fact about a quarter of the electricity generated in the United States, for example, is used to heat space or water. This would bring the correction factor down from 3 to 2.25. This quality ratio would also change if one were to run entire economies on electricity derived from renewable resources to avoid carbonaceous fuels entirely (e.g. Heinberg and Fridley 2016). If this were the case then probably a majority of uses would be for uses that do not require electricity, so the quality factor should be applied to only some part of it, such as trains with high efficiency electric motors (e.g. Freidemann 2015). Thus some of our earlier studies were examining energy sources to entire economies from this perspective. Obviously it is a complex issue and one needs to ask "what exactly am I analyzing and for what purpose?" (Carbajalis Dale et al. 2015).

11.7.5 Theoretical Versus Empirical Assessments of Electricity Output

Some earlier studies apparently used "nameplate" values (1,800 kWh/M^2-yr) for assessing electricity outputs from PV facilities or estimates from the "IEA Guidelines" that assume 1,700 kWh/M^2-yr, a value not appropriate for a Northern, cloudy country (The guidelines do suggest a reduction for real conditions). In fact the actual output is a tricky business: very often clouds, bird droppings, overheating, dust accumulation, lightning storms, equipment failures and degradations over time can decrease actual output compared to "nameplate" values. Additionally too much

11.7 Methodological Issues, Problems and Criticisms of EROI

output can fry electrical components at various locations in the grid. Prieto and Hall found that the actual output for a facility in Spain with a nominal output of 1,800 kWh/M^2-yr was a measured 1,375 kWh/M^2-yr. Ferroni and Hopkirk (2016) likewise found measured values considerably less than nameplate values.

A related issue concerns the assumptions about how long the facility will last. Most investigators have applied a life span of 25 years for PV facilities, and the IEA guidelines suggest 30 years. Since most solar facilities are new this is hard to measure, but in fact in reality it may be less. Ferroni and Hopkirk (2016) came up with an estimate of a mean of 18 years for Switzerland. Prieto (personal communication) believes it is much less than 25 years in Spain because many companies have declared bankruptcy and thus do not honor their warranties. Without warranties or the specific parts to fix failures many PV facilities in Spain have been abandoned or completely refigured.

11.7.6 Boundaries and Comprehensiveness of the Cost Assessments

Carbajalis Dale et al. (2015) state in a footnote in reference to Prieto and Hall's study: "Though arguably, somewhat inconsistent in its definition of system boundary and arbitrary in its inclusion of a large number of nonenergy inputs." Likewise, in the on line "Commodities and Future trading" "Renewables have a higher EROI than fossil fuels" we can read "(Prieto and Hall) add every incidental energy cost they can think of, like the energy costs of building fences around the solar farm, and so on. They even add energy costs for things like corporate management, security, taxes, fairs, exhibitions, notary public fees, accountants, and so on (monetary costs are converted into energy by means of a formula)". We respond "As if these were not legitimate energy costs to build and operate PV plants?" In fact they are, for example given the high value of e.g. scrap copper such plants are very susceptible to thieves stealing electrical components (a cost, incidentally, Prieto and Hall included). Thus fences and security systems are needed or the EROI goes to zero. So are roads, module washing and financial institutions. Based on the earlier studies of e.g. Hannon (1981) it is clear that all services (as well as goods) require substantial amounts of energy (roughly a third to half per dollar compared to societal means) to undertake. In order to get a comprehensive assessment we "followed the money" and assigned a (conservative, we believe...one third the national mean) energy cost to all service expenditures, which Prieto had a good estimate of because as chief site engineer he signed for every item and activity associated with building a gigawatt plant in Spain. We assume all of the services we mentioned are not incidental but necessary and should be included in energy costs, and have not heard any explicit reason why we should exclude any one.

Prieto and Hall found that the construction of modules and basic electronic components such as inverters were only about a third of the total energy cost of actually building and operating a solar facility in Spain. We assume the same applies elsewhere, but have not seen from any of our critics an analysis that includes any large part of the real energy costs that we did.

Raugei et al. (2013) have argued that Prieto and Hall include for PV various costs, such as site preparation and various environmental issues, that are not included in our assessments of oil or coal. This is not true, as all such costs are included (in theory) with our indirect energy assessments which are based on total "upstream" expenditures by industries. We agree with them that the boundaries should include all energy costs that any energy gathering activity experinces.

11.7.7 Technological Changes Over Time

Another issue raised by Raugei and other solar analysts is that the monetary and presumably the energy cost of making solar PV modules has been declining for decades, and is expected to continue to do so, although perhaps at a declining rate. Thus they criticize the Prieto and Hall study for using technology appropriate for 2008 (actually we used 2009–2011 technology) when there has been a perhaps 10–20 % decline in energy cost of making modules since then (some of which, in terms of money if not energy, is attributable to e.g. subsidies of the process by the Chinese government). To our knowledge there has not been a similar decline in other inputs to PV systems. I agree with them, and believe that one should do costs and benefits for particular, given years.

Swenson (2016) argues that if solar (PV) technologies are adapted at large scale the entire system will be forced to become much more efficient than fossil fueled systems and can exist with a lower EROI. As empiricists we cannot comment on this. But we add that the greater proportion of our total energy provided by "renewable" sources the greater the costs that will be required for storage and integration with the grid.

11.8 Other Issues That Need Consideration That Might Decrease the EROI of Solar Energy: Storage

A large issue pertaining to electricity derived directly from present solar is storage. Since sunshine and, especially, wind is dependent upon nature's only partially predictable whims, and cannot be programmed in advance, meeting the demand load can be very difficult. A day might be sunny or cloudy (with half or less of the insolation), and wind blows on average only about 30 % of the time (nearer 20 % in Germany), and there may be periods of two weeks or more with no wind at all. The supply of power from photovoltaic systems is only somewhat more predictable. Additional energy expenses are required to compensate for these intermittencies.

Two basic strategies are required: the first is storage, the second some other kind of readily dispatchable power. While storage in automobile batteries is very familiar to us, all of the batteries in the world would store less than one minute of global electrical output, and it is not cost effective on a massive scale. There are many possibilities to try to change this: massive investments in new lithium batteries, giant flywheels, pressurized underground chambers and generation of hydrogen are a few, but the only one that is feasible on a large scale today is elevated storage of water, whether in existing hydroelectric facilities or in specially constructed pumped storage facilities. Some problems with using these systems, which can respond very quickly, are: (1) there is an electricity loss of some 25–35 % in the pump up and later release systems (2) the availability of such sites on rivers are limited and (3) the intermittent release of water can play havoc with fish and aquatic ecosystems (Ward and Stanford 1979). Pumped storage systems on lakes have fewer such problems. For example the giant Ludington water storage project makes the use of renewable energy much more feasible in Michigan. Also in the Great Lakes region long transmission lines allow "load sharing" as wind may blow at different places at different rates: the larger the area the more different facilities can cover each other. In practice increasingly all of these difficulties in following loads with intermittent sources are dealt with through the use of gas turbines, which are essentially natural gas powered jet engines mounted on pedestals and connected with generators, which can be turned on and off relatively quickly. Since modern gas turbines can be up to 50 % efficient (vs. 30–40 % for a conventional coal or oil plant) by using the waste heat from the turbine to run a steam turbine, they are efficient and cheap. One down side is that changing the speed of the turbines frequently causes much greater wear. While it is true that base load systems, such as nuclear or coal, are also difficult to increase or decrease to follow the load, any substantial penetration of intermittent energy into the grid will probably be very energy costly to accommodate to. But for the present discussion there has been little assessment of how much energy such systems would cost, or how much energy cost (probably large) should be added to EROI assessments. One exception is that Carbajalis Dale et al. (2014) estimate that adding a relatively small amount of storage to PV systems would quickly put them into energy deficit. On a smaller scale Palmer (2013) found that batteries doubled the energy cost of rooftop solar systems.

These energy costs tend to be ignored by PV and wind advocates, who also argue that coal and nuclear facilities have their own problems with responding to variable loads (which, however, are being met readily now). In future assessments these costs (probably large) should be added to all EROI assessments, and they are likely to be very large if we expand intermittent renewables as much as many advocate.

11.9 Exponential Growth of Energy Production

Many advocate that we must grow renewable systems very rapidly, continuing indefinitely the exponential growth seen in the early part of the twenty-first century. Several investigators have calculated the EROI and energy delivered to society of an exponentially growing PV industry while assuming that new modules are generated from the output of existing similar power plants. Neumeyer and Goldston (2016) found that an initial EROI of 10:1 quickly dropped to 2:1 as most of the power output went to generating new plants. Carbajalis Dale and Benson (2013) and Kaufmann and Shiers (2008) found a similar very sharp drop in net power output if growth were large. Thus if we are to have a large, exponentially growing PV system the new facilities almost certainly will have to be constructed using fossil fuels. Another issue is whether there would be enough materials to do this. Fizaine and Court (2015) and Gupta and Hall (2012) found that an exponentially growing PV system might run out of copper in a very few decades. Hertwich et al. (2014) found that photovoltaic systems can use 11–40 times more copper than conventional fossil generation systems, although they also thought that there was enough copper to build a large renewable system.

11.10 Summary of Critiques

Many of the various critics' points have been answered somewhere in the literature (and of course are covered in the previous pages) although not always in all pertinent publications. But what we do not yet have is an objective summary of what values and boundaries should in fact be used, because most assessments are from investigators with a particular perspective. Finally it is my perspective that most (but not all) of the remaining criticisms have been or can be dealt with, at least through sensitivity analysis, leaving us with a more robust set of tools and with some good projects for future graduate students. What is more difficult is that it takes time and money to undertake these assessments, and at least so far there has been little interest by funding agencies to support such studies. This is another good reason that we need a well-funded, objective, international program for undertaking and comparing EROI, including a good program of data gathering. This appears to be changing in the U. S. The National Science Foundation, which has (gracefully) acknowledged its irresponsibility in not having funded such a program earlier, is now developing such a program (Love and Murphy 2016). One can only hope.

11.11 Further Issues in Comprehensiveness of EROI Analysis

Frequently, as we have seen above, the most important reasons for differences between different analyses are the boundaries used for estimating costs. While it might seem straightforward to decide what the boundaries should be in fact it is quite difficult because the inclusion of many of the costs can be more philosophical than scientific.

11.12 Business Services and Taxes

All investigators agree that one should include the direct (on site) and the most obvious indirect energy costs (Murphy et al. 2011). But what about the energy used to support the business services used, as business services require energy-requiring brick and mortar buildings, which have to be heated, cooled, supplied with lighting and electricity etc.? Or should there be an assessment of the energy used to support the taxes paid? Taxes, when spent, require energy to support them too—such as the energy to build and maintain roads, provide schooling and so on. One might argue that an oil field or gas field in Pennsylvania or North Dakota or a PV facility in Spain requires considerable construction and maintenance costs that are paid for by governments that in turn are operating off taxes. For example Pennsylvania has found that there are very high costs associated with the new "fracked" gas wells due to the heavy trucks full of water driven over low quality roads during all seasons of the year, creating much damage that has to be fixed with heavy equipment. In addition the driller's children require schooling and there is a general increase in the need for policing and health services (e.g. Dutzik et al. 2012; Food & Water watch 2013 and references therein). I believe, therefore, that tax expenditures, and the energy required to generate those governmental services, are properly part of the energy cost of a project.

11.13 Labor

Perhaps most controversial is whether to include the energy required to support labor. One can visualize a number of levels of inclusion. Certainly laborers use energy, and since a hard working person might use some 1.8 MJ/h this might be included. Since a person is a machine that works at about 20 % efficiency then the food energy required would be 9 MJ/h. This would not be large compared to the machines most laborers would be using, say a diesel engine using 4 L (135 MJ) per hour. But labor would not be available if it were not paid, and one might want to include the energy required to support the worker's paycheck. Assume a worker is paid $50,000 a year. Energy must be spent within the economy to produce the goods and services demanded by the worker or his/her family spending the paycheck. In 2015 the United States Economy used roughly 5.6 MJ per average dollar of GDP. Thus, assuming that our worker's family spends their money on "average" goods and services, it would take about 280,000 MJs of energy (equal to 46 barrels of oil) to support his or her paycheck. When we were first considering EROI, back in about 1970, we presented this concept to some economists. They said it was inappropriate as that was *consumption*, which should not be added to *production*. Well I still think the energy to support the worker's paychecks is legitimately part of the cost of production, but it is so controversial that we do not use it except occasionally in sensitivity analysis.

Summary

Thus I conclude that one should include all real costs that are used to generate and operate (for example real energy costs of access roads, finance and taxes, all operation and maintenance for an energy generation facility). Most analyses to date have not included energy used beyond the "direct and indirect" energy costs (and usually not all of them) and as such produce low energy costs and hence higher EROIs than is probably the actual case. Regrettably, a clear definition of system boundaries is missing in many published EROI studies of electricity production. However these costs are relatively small and should be included by using financial costs and energy intensities. We need much better governmental support and more explicit and open peer review to get genuinely reliable results. The question as to whether or how one should include labor remains unresolved.

Acknowledgments I thank Carey King, Pedro Prieto and Marco Raugei for discussion and clarification on this section, although the final product remains my own.

References

Arvesen, A., and E.G. Hertwich. 2015. More caution is needed when using life cycle assessment to determine energy return on investment (EROI). *Energy Policy* 76: 1–6.

Brandt, A.R., M. Carbajalis Dale, and C. Barnhart. 2013. Calculating systems-scale energy efficiency and energy returns: a bottom-up matrix-based approach. *Energy* 62: 235–247.

Bullard, C.W., B. Hannon, and R.A. Herendeen. 1975. *Energy flow through the US Economy*. Urbana: University of Illinois Press.

Carbajalis Dale, M., and S.M. Benson. 2013. The energy balance of the photovoltaic (PV) industry - is the PV industry a net energy provider? *Environmental Science and Technology* 47(7): 3482–3489.

Carbajalis Dale, M., M. Raugei, C.J. Barnhart, and V. Fthenakis. 2015. Energy return on investment (EROI) of solar PV: an attempt at reconciliation. *Proceedings of the IEEE*. doi:10.1109/JPROC.2015.2438471.

Cleveland, C.J. Energy quality. In: Cutler J. Cleveland, Editor-in-Chief, The Encyclopedia of Earth. Published by The Environmental Information Coalition, National Council for Science and the Environment, Washington, D.C.

Cleveland, C.J., D.I. Stern, and R.K. Kaufmann. 2000. Aggregation and the Role of Energy in the Economy. *Ecological Economics* 32: 301–317.

Cleveland, C.J. 2005. Net energy from oil and gas extraction in the United States, 1954-1997. *Energy* 30: 769–782.

Cleveland, C.J., R. Costanza, C.A.S. Hall, and R. Kaufmann. 1984. Energy and the United States economy: a biophysical perspective. *Science* 225: 890–897.

Carbajalis Dale, M., C.J. Barnhart, and S.M. Benson. 2014. Can we afford storage? A dynamic net energy analysis of renewable electricity generation firmed by energy storage. *Energy and Environmental Science* 7(5): 1538–1544.

Dutzik, T., E. Ridlington and J. Rumpler, 2012. The Costs of Fracking. The Price Tag of Dirty Drilling's Environmental Damage. PennEnvironment, Research & Policy Center. Philadelphia.

Fthenakis V.M. and Kim H.C., 2011. Photovoltaics: Life -cycle analyses. *Solar Energy*: 85: 1609–1628.

Food&Water Watch. 2013. The social costs of fracking: A Pennsylvania Case study. Food&Water Watch, Washington D.C.

References

Ferroni, F., and R.J. Hopkirk. 2016. Energy Return on Energy Invested (ERoEI) for photovoltaic solar systems in regions of moderate insolation. *Energy Policy* 94: 336–344.

Freidemann, A. 2015. *When trucks stop running*. Springer, N.Y.

Fthenakis, V. R. Frischknecht, M. Raugei, H. C. Kim, E. Alsema, M. Held and M. de. Wild-Scholten, 2011, Methodology Guidelines on Life Cycle Assessment of Photovoltaic Electricity, 2nd edition, IEA PVPS Task 12, International Energy Agency Photovoltaic Power systems Programme.

Fizaine, F., and V. Court. 2015. Renewable electricity producing technologies and metal depletion: A sensitivity analysis using the EROI. *Ecological Economics* 110: 106–118.

Fthenakis V.M. and Kim H.C., 2011. Photovoltaics: life-cycle analyses. *Solar Energy* 1609–1628.

Grandell, L., C.A.S., Hall, and M. Hook. 2011 Energy return on investment for Norwegian oil and gas in 1991-2008: Sustainability: Special Issue on EROI. 2011. pp. 2050–2070.

Guilford, M., C.A.S. Hall, P. O'Conner, C.J. Cleveland (2011) A new long term assessment of EROI for U.S. oil and gas. Sustainability: Special Issue on EROI. 1866–1887.

Gupta, A.K., and C.A.S. Hall. 2012. Energy cost of materials: materials for thin-film. In *Fundamentals of Materials for Energy and Environmental Sustainability*, ed. D.S. Ginley, and D. Cahen. Cambridge: Cambridge Univ. Press.

Hall, C.A.S. 2015. EROI and its implications for long-term prosperity. In Research methods and applications in environmental studies, ed. Ruth, M., 197–224. Cheltenham, England: Edward Elgar, 534pp.

Hall, C.A.S. and J.W. Day (eds.) 1977. Ecosystem modeling in theory and practice. An introduction with case histories. Wiley Interscience, NY. 684 pp.

Hall, C.A.S., M. Lavine, and J. Sloane. 1979. Efficiency of energy delivery systems: Part I. An economic and energy analysis. Environmental Management 3(6): 493–504.

Hall, C.A.S., and C.J. Cleveland. 1981. Petroleum drilling and production in the United States: Yield per effort and net energy analysis. *Science* 211: 576–579.

Hall, C.A.S., C.J. Cleveland, and R. Kaufmann. 1986. *Energy and Resource Quality: The ecology of the economic process*. NY: Wiley Interscience.

Hall, C.A.S., S. Balogh, and D.J.R. Murphy. 2009. What is the Minimum EROI that a Sustainable Society Must Have? *Energies* 2: 25–47.

Hall, C.A.S., B. Carbajalis Dale, and D. Pimentel. 2011. Seeking to understand the reasons for different estimates of the EROI for biofuels. *Sustainability* 2011(3): 2433–2442.

Hall, C., A.S. J., G. Lambert, S.B. Balogh. 2014. EROI of different fuels and the implications for society. *Energy Policy Energy Policy* 64: 141–152.

Hannon B. 1981. Analysis of the energy cost of economic activities: 1963 2000. Energy Research Group Doc. No. 316. Urbana: University of Illinois.

Herendeen, R., and C. Bullard. 1975. The energy costs of Goods and Services. 1963 and 1967. *Energy Policy* 3: 268.

Heinberg, R., and D. Fridley. 2016. *Our renewable future: laying the path for one hundred percent clean energy*. Washington, D.C.: Island Press.

Hertwich, T.G., E.A. Boumana, A. Arvesena, S. Suhb, G.A. Heath, J.D. Bergesenb, A. Ramirez, M.I. Vegae, and L. Shif. 2014. Integrated life-cycle assessment of electricity-supply scenarios confirms global environmental benefit of low-carbon technologies Proc. *Natl. Acad. Sci. USA* 112(20): 6277–6282.

Hopkinson, C.S., and J.W. Day. 1980. Net energy analysis of alcohol production from sugarcane. *Sci* 207(4428): 302–304.

Hu, Y., C.A.S. Hall, J. Wang, L. Feng, and A. Poisson. 2013. Energy Return on Investment (EROI) of China's conventional fossil fuels: Historical and future trends. *Energy* 54: 352–364.

Kaufmann, R. Chapter 8 in Hall, C.A.S., C.J. Cleveland and R. Kaufmann. 1986. *Energy and resource quality: the ecology of the economic process*. Wiley Interscience, NY. 577 pp.

Kaufmann, R.K., and L.D. Shiers. 2008. Alternatives to conventional crude oil: When, how quickly, and market driven? *Ecological Economics* 67: 405–411.

King, C.W., J.P. Maxwell, and A. Donovan. 2015. Comparing world economic and net energy metrics, Part 1: Single Technology and Commodity Perspective. *Energies* 8: 12949–12974.

Kubiszewski, I., C. Cleveland, and P. Endres. 2010. Meta-analysis of net energy return for wind power systems. *Renewable Energy* 2010(35): 218–225.

Lambert, J., C.A.S. Hall and S. Balogh. 2013. *EROI of global energy resources: status, trends and social implications*. Report to Division of Foreign Investment, United Kingdom. 136 pp.

Lambert, J., C.A.S. Hall, S. Balogh, A. Gupta, and M. Arnold. 2014. Energy, EROI and quality of life. *Energy Policy* 64: 153–167.

Leccisi, E., M. Raugei, and V. Fthenakis. 2016. The energy and environmental performance of ground-mounted photovoltaic systems – a timely update. *Energies* 9(8): 622. doi:10.3390/en9080622.

Lenzen, M. 2008. Life cycle energy and greenhouse gas emissions of nuclear energy: A review. *Energy Conversion and Management* 49: 2178–2199.

Love, T., and D. Murphy. 2016. White Paper: Implications of Net Energy for the Food-Energy-Water Nexus An NSF-funded workshop at Linfield College, McMinnville, OR 14-16 January 2016 Award Number: 1541988.

Mulder, K., and N.J. Hagens. 2008. Energy Return on Investment: Toward a Consistent Framework. *Ambio* 37: 74–79.

Murphy, D.J., and C.A.S. Hall. 2010. Year in review—EROI or energy return on (energy) invested. Annals of the New York Academy of Sciences. *Special Issue Ecological Economics Reviews* 1185: 102–118.

Murphy, D., Hall, C.A.S., Cleveland, C., and P. O'Conner. 2011. Order from chaos: a preliminary protocol for determining EROI for fuels. Sustainability: Special Issue on EROI, pp. 1888–1907.

Murphy, D., and Carbajalis Dale, M. 2016. Comparing Apples to Apples: Why the Net Energy Analysis community needs to adopt the LCA framework. Energies, in press.

Neumeyer, C., and R. Goldston. 2016. Dynamic EROI Assessment of the IPCC 21st Century Electricity Production Scenario. *Sustainability* 8(5): 421.

Palmer, G. 2013. Household solar photovoltaics: supplier of marginal abatement, or primary source of low-emission power? *Sustainability* 5(4): 1406–1442.

Palmer, G. 2014. *Energy in Australia: peak oil, solar power, and Asia's economic growth*. N.Y.: Springer Briefs in Energy.

Pracha, A.S., and T.A. Volk. 2011. An edible energy return on investment (EROI) analysis for wheat and rice in Pakistan. *Sustainability* 3: 2358–2391.

Prieto, P.A., and C. Hall. 2013. *Spain's photovoltaic revolution: the energy return on investment*. N.Y.: Springer Briefs in Energy.

Patzek, T. 2004. Thermodynamics of the corn-ethanol biofuel cycle. *Critical Review in Plant Sciences* 23: 519–567.

Patzek, T.W. 2007. A first-law thermodynamic analysis of the corn-ethanol cycle. *Natural Resources Research* 15: 255–270.

Pimentel, D. 2003. Ethanol fuels: energy balance, economics, and environmental impacts are negative. *Natural Resources Research* 12: 127–134.

Raugei, M., Fullana-i-Palmer and V. Fthenakis. 2012. The energy return on energy investment (EROI) of photovoltaics: Methodology and comparisons with fossil fuel life cycles. *Energy Policy* 45: 576–582.

Raugei, M. 2013. Comments on "Energy intensities, EROIs (energy returned on invested), and energy payback times of electricity generating power plants"-making clear of quite some confusion. *Energy* 59: 781.

Raugei, M., M. Carbajalis Dale, C.J. Barnhart, and V. Fthenakis. 2015. Rebuttal: "Comments on 'Energy intensities, EROIs (energy returned on invested), and energy payback times of electricity generating power plants' - Making clear of quite some confusion". *Energy* 82(15): 1088–1091.

Raugei M., and Leccisi E., 2016. A comprehensive assessment of the energy performance of the full range of electricity generation technologies deployed in the United Kingdom. *Energy Policy* 90: 46–59.

References

Swenson, R. 2016. The solar evolution: Much More with Way Less, Right Now—The Disruptive Shift to Renewables. *Energies* 2016(9): 676.

U. S. Department of Energy, 2000; U.S. Energy Information Agency, various years.

U.S. Bureau of Economic Analysis, Department of Census, various years.

United States Geological Survey (USGS) (2003), *The World Petroleum Assessment 2000* www.usgs.gov.

U.S. Energy Information Agency (2016) "Definitions of Petroleum Products and Other Terms (Revised May 2010)" and Natural Gas "Appendix A Explanatory notes".

Ward and Stanford. 1979. *The ecology of regulated streams*. New York: Plenum Press.

Weissbach, D., G.A. Ruprecht, K. Huke, S. Czerski, and A. Hussein Gottlieb. 2013. Energy intensities, EROIs (energy returned on invested), and energy payback times of electricity generating power plants. *Energy* 52: 210–221.

Weissbach, D., G.A. Ruprecht, K. Huke, S. Czerski, and A. Hussein Gottlieb. 2014. Reply on "Comments on 'Energy intensities, EROIs (energy returned on invested), and energy payback times of electricity generating power plants' – Making clear of quite some confusion. *Energy* 68: 1004–1006.

Chapter 12
The History, Future, and Implications of EROI for Society

12.1 Sustainability

Perhaps the most important issue facing mankind is whether or not our civilization is sustainable (e.g., Heinberg and Fridley 2016). While most people do not think much about this topic, there certainly is enough information and literature to make us understand that this is a real issue of potential enormous importance. Will our civilization, country, even species be here in several decades? Centuries? Millennia? Indefinitely? There are many suggestions about how to make this more probable, i.e., more sustainable, but the vast majority of them examine only a very small part of the overall issue. Part of the problem is that sustainability has been defined in various ways according to the interests and foci of the various defining groups. According to an important paper by Goodland and Daly (1996) the different groups, which they call environmental, social, and economic, generally have little knowledge or interest in the viewpoints of other groups and are often, in fact, contradictory. For example, attempting to ensure the sustainability of a particular group or culture might be undertaken through economic growth, which might make the group or region less sustainable environmentally over the longer time. (From our perspective "environmental" would mean resource as well as pollutional/degradational issues.)

Over the longer time many, perhaps most, civilizations in fact have not been sustainable. Some 99 % of species that have ever existed on Earth are extinct, and likewise most of the great civilizations of the past went extinct, or something like it, too (Tainter 1988; Diamond 2005). Good energy investments, in both biological and human economic systems, favor but do not guarantee survival. Life does not come complete with guarantees. In fact, as we have seen in Chap. 6, life depends on natural selection, and natural selection required failures as well as successes. Nevertheless most people act as if they want their particular civilization, or some part of it, to survive and prosper indefinitely. Our question is "what does EROI have to do with this?"

Since we have become such an energy-dependent society, a critical aspect of sustainability, however defined, is ensuring that there is sufficient energy for maintenance metabolism of that society, or, as many economists would argue, growth of the economy. There are a number of biophysical reasons to be concerned about our potential for sustainability, including depletion of our most important fuels, possible climate impacts of continuing their use, and difficulties in getting alternative energies, either renewable, nuclear or something else, when the fossil fuels are essentially gone or too expensive. In all of this EROI is an important, sometimes critical issue, and that is what this chapter is about.

12.2 Peak Oil: How Long Can We Depend on Oil and Other Fossil Fuels?

It does not matter what the EROI of a fuel is if it is depleted. So, we first examine the issue of how much fossil fuels we are likely to have. The best-known model of oil production was developed by Marion King Hubbert, who proposed that the discovery and production of petroleum over time would first grow exponentially and then reach a peak when roughly half the resource had been exploited, followed by a more or less symmetrical decline; that is the production of oil for a region, a country or the world would follow a more or less symmetric, bell-shaped curve. He surmised that the rate of production would initially increase exponentially as the means for exploiting and using the resource were figured out. A peak in production would occur when about 50 % of the ultimate recoverable resources (URR) had been extracted (he later opined that there may be more than one peak). URR is the total amount of oil that can be recovered. Hubbert's hypothesis was based principally on his experience as a geologist and it was not a bad guess. He famously predicted in 1956 that US oil production from the lower 48 states would peak in 1970, which in fact it did. Hubbert also predicted that the US production of natural gas would peak in about 1980, which it did, although it has since shown signs of recovery and there is a second peak in early 2015 based on large part on "unconventional" and "shale gas." He also predicted that world oil production would peak in about 2000 (see Bartlett 2000). In fact, the production of oil globally continued to increase until 2005, after which it appears to have entered an "undulating plateau," as predicted earlier by geologists Campbell and Laherrere (1998; Fig. 12.1). The main reason that the Hubbert Curve "works" is that if the finding of oil peaks at some point then inevitably and due to the simplest of math (the integral of the production curve cannot exceed the integral of the finding curve) the peak of production must follow. For the United States the peak in finding oil was in the 1930s and for the world in the 1960s. Both have declined enormously since then and production declines must follow (Fig. 12.2).

In the past several decades, a number of 'neo-Hubbertarians' have made predictions about the timing of global and national oil production peaks using several variations of Hubbert's approach. Various forecasts of the timing of global peak have ranged from one predicted for 1989 to many predicted for the first decade of the Twenty-first century to one as late as 2040. The difference in these forecasts is largely due to differences in assumptions about the URR (ultimate recoverable

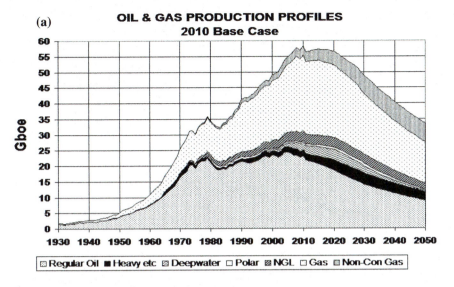

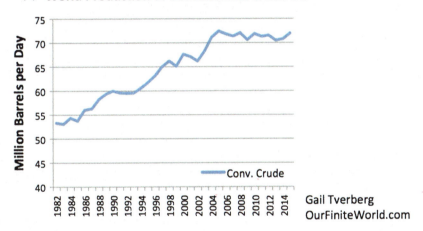

Fig. 12.1 a Rate of production of oil and other hydrocarbons for the world. The lower four categories are what is normally considered "oil," which has been at an "undulating plateau" since about 2005. NGL are natural gas liquids, produced along with natural gas which has a later peak. (*Courtesy* Colin Campbell) **b** Global production of conventional oil (*Courtesy* Gail Tverberg Our Finite World)

resource). Brandt (2007), Nashawi et al. (2010) and Hallock et al. (2014) show that the Hubbert Curve is a good predictor for most post-peak nations, which includes the majority of all oil-producing nations. Other forecasts for world oil production do not rely on such curve-fitting techniques to make future projections and/or *a prior* assumption about URR. According to one forecast by the US Energy Information Agency (EIA 2003), world oil supply in 2025 will exceed the 2001 level by 53 %.

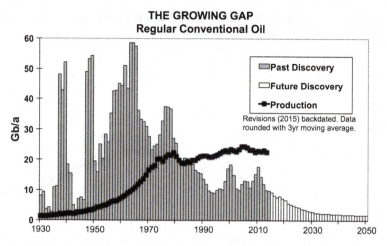

Fig. 12.2 Global rate of finding (*vertical bars*) and producing (*dark squares*) conventional oil. Note that our rate of finding oil peaked in the 1960s, but that the use rate increased over time until about 2015. Thus, as Hubbert predicted, the peak in finding allows one to project that a peak in production must follow (image courtesy of Colin Campbell). According to Holter (2016) discoveries in 2016 were very small compared to earlier in the decade

The EIA reviewed five other world oil models and found that all of them predict that production will increase over the next two decades to around 100 million barrels per day, substantially more than the 80 million barrels per day produced in 2016 (The production of "all liquids", which includes conventional crude oil plus fracked oil, natural gas liquids, biofuels, refinery gains and so on, was 96 million barrels per day). Several of these models rely on the 2000 USGS estimates of URR for oil. Thus, it does seem that the production of "conventional oil" is at a rough plateau, but that other materials such as natural gas liquids and shale oil are still increasing. At a minimum, the annual rate of increase in global oil production has fallen from about 5 % to less than 1 % (Fig. 12.3).

Hallock et al. (2004, 2014) undertook a thorough analysis of USGS estimates of URR by modeling a Hubbert approach to future production of conventional oil based on the USGS low, medium (best), and high estimates of URR. They published their first analysis in 2004 where they made projections of future oil production for the 46 largest oil-producing nations using a Hubbert curve model and the USGS low, medium, and high estimates of URR. Then in 2014 they returned to their predictions made in 2004 and assessed the actual production of all 46 countries versus their earlier modeled predictions (Fig. 12.4). The found that the majority (30) of nations, with about 75 % of all oil production, had a production trajectory that closely approximated their earlier prediction that had assumed a low URR, some countries (7, with about 12 % of production) followed a trajectory consistent with the medium URR, and 8 (with a production of about 8 % of total) followed a trajectory consistent with medium to high URRs. One country had a very erratic pattern that could not be characterized. This seems to be very solid support for the

12.2 Peak Oil: How Long Can We Depend … 149

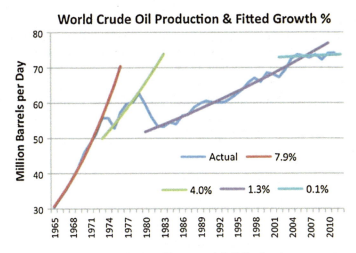

Fig. 12.3 Growth of global oil production (Courtesy of Gail Tverberg)

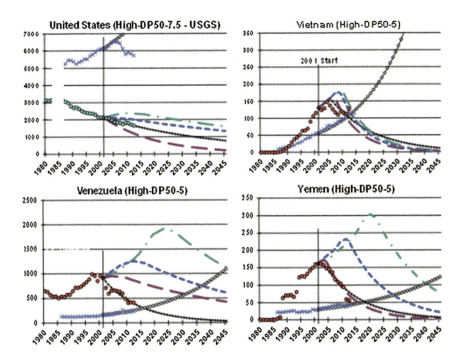

Fig. 12.4 Some typical results from Hallock et al. (2014) for conventional oil, where their predictions from a decade earlier (at *vertical gray line*) are examined against actual production (colored circles; purple, blue and aqua intermittent lines are simulations made in 2004 for low, medium and high EURs; blue crosses are measured consumption followed by extrapolations). These graphs also show how many countries are following pretty typical Hubbert curves. For other countries, most of which are more or less similar, see: http://www.sciencedirect.com/science/journal/03605442/64/supp/C; http://www.sciencedirect.com/science/journal/03605442/open-access

assumption that the actual amount of producible oil is nearer the low or between the medium and low estimates of USGS and not near the high. Observed (empirical) oil production profiles given in the appendix of that paper show for the most part very clear Hubbert patterns of production. In my opinion, this paper deserves a lot more attention than it has had.

It should be noted that the majority of oil-supply forecasts, with the exception of Hallock et al. (2004, 2014) and Nashawi et al. (2010) have had a poor or ambiguous track record, regardless of method. Most recent results of curve-fitting methods showed a consistent tendency to predict a peak within a few years, and then a decline, no matter when the predictions were made. It is now a well-established fact that economic and institutional factors, as well as geology, were responsible for the US peak in production in 1970, forces that are explicitly excluded from the curve-fitting models (Kaufmann and Cleveland 2001). Thus, the ability (or the luck) of Hubbert's model (and its variants) to forecast production in the 48 lower states accurately cannot necessarily be extrapolated to other regions or the world. It is too early to tell. But the evidence so far is that while the Hubbert curve might not predict the date of peak oil for the world perfectly, the general concept is very sound for most oil-producing countries and future researchers will look at our present uncertainty as small blips compared to the overall pattern of depletion over time. The production of conventional oil globally is decreasing or barely increasing despite enormous investments made in the last decade (Hallock et al. 2014; Fig. 12.2). The principal reason is that most oil production comes from very large oil fields (greater than 500 million barrels, called "elephants") and we have found few elephants in the world since the 1960s. Now these large oil fields are aging, and their production is falling (Simmons 2002), as are their EROIs (Tripathi and Brandt, in revision). It appears exceedingly unlikely that new oil discoveries, most of which are not large, will make up for the decline in the elephants. There has been some increase lately in other liquids, such as natural gas liquids, but little or none in real conventional oil.

In past decades many of us thought that oil would likely come into short supply more or less by now, but that gas and especially coal would last much longer. The most recent analyses we have (Maggio and Cacciola 2012; Mohr et al. 2015; Laherrere 2016), predict that we will have a peak in ALL fossil fuels by from 2025 to 2050 (see Fig. 9.5). While the rate of growth of demand for oil by importing countries has eased off in recent years there is a new concern: that the internal consumption of oil by oil-producing nations has tended to increase so that there is much less oil for export, and this trend is likely to continue (e.g., Hallock et al. 2004; A good up to date record is maintained on the web site of Matt Mushalik in Australia).

12.3 New Technologies to the Rescue?

It is possible that a production peak could come earlier, based on human efforts to reduce CO_2 emissions (McGlade and Ekins 2015). Energy authority Vaclav Smil (2011) has summarized the challenges of moving off mostly fossil fuels:

12.3 New Technologies to the Rescue?

There are five major reasons that the transition from fossil to nonfossil supply will be much more difficult than is commonly realized: scale of the shift; lower energy density of replacement fuels; substantially lower power density of renewable energy extraction; intermittence of renewable flows; and uneven distribution of renewable energy resources.

Whether renewable energies, such as wind, biomass and solar PV, could replace some large part of the fossil fuels anytime soon seems to be highly unlikely, although advocates suggest that it is possible (Jacobson and Delucchi 2009; Jacobson et al. 2015; Heinberg and Fridley 2016; see Hall 2016). In "Our renewable future" Heinberg and Fridley believe the need to avoid climate catastrophe will make the investment worth it even at the enormous cost it would entail (estimated on page 123 as 20 times the present rate of all investments in renewables for many decades). Much of the rest of the book delineates the possibilities and difficulties with each step of this proposed transformation. David MacKay (2010) concluded: "we must have no delusions about the area required for large-scale solar power; about the challenge of transmitting energy over large distances; about the additional costs of handling intermittency; and about the need for breakthroughs not only in the whole-system costs of photovoltaics but also in the cost of systems for storing energy. CSP (concentrating solar) plants need to be in safe locations, and the ultra-high voltage direct current transmission (UHVDC) system required in order to transport the electricity to points of final use. This is not currently feasible in North Africa, for example". Trainer (2013) likewise found that costs of renewables were extremely large.

To give an example of the difficulties, today most renewable energy comes from hydropower and biomass, and the contribution of the latter is declining, so that the total contribution from all renewable in the U.S. has barely increased from 11 % in 2010 to 12.6 % in 2016 to a projected 16.1 % by 2040 (U.S. EIA). Meanwhile, the EIA projects that all fossil fuels will continue to increase in absolute terms. So much for reducing CO_2 emissions! Some solar advocates project a much higher transformation rate, in line with what these authors see as necessary, as prices for, e.g., PV-generated electricity decline. We shall see. Meanwhile the world's economy has essentially stopped growing, perhaps in response to increasing resource limitations, implying a very different future that might greatly change our projections and options.

There is a new player on the block, however, one that gives more than a little credibility to the arguments of economists like Adelman and Lynch (1997) that higher oil prices will stimulate new technologies. The concept of fracking is an old one, and has been occasionally used to enhance oil production as long ago as about 1930. Likewise horizontal drilling has been around for a very long time. What is new is their combination along with the use of special sand to prop open the cracks made by high-pressure water sent down hole (Fig. 4.6). This new technological package has been used to exploit extensive horizontal source beds of oil in Texas, North Dakota, and elsewhere. While there is a lot of oil in these formations, and it is often of good quality, the formations are characterized by low porosity and low permeability. As a result production in these wells tends to decline very rapidly compared to conventional oil wells. On the plus side there are few dry holes

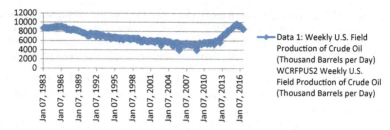

Fig. 12.5 Time series of oil production in the United States 1983–2016/The peak in US oil production of 10,500 thousand barrels per day occurred in 1970 (*source* U.S. EIA; http://www.eia.gov/dnav/pet/hist/LeafHandler.ashx?n=PET&s=WCRFPUS2&f=W)

because the oil is in massive, relatively uniform formations rather than more specific traps. The result has been a dramatic reversal in the long-term decline of oil production in the United States (Fig. 12.5). A peak nearly at the level of the 1970 peak was achieved in 2015, although production since then has been declining, whether due to the depletion of the "sweet spots" or, as is likely, the decline in effort due to the low price of oil. In 2016, the US again began importing about half the oil it consumes. It is unlikely that fracking will do other than delay the inevitable U.S. peak and decline by more than about a decade. On the other hand it seems that so far every time the US is about to have a catastrophic decline in oil production something comes on line to give us at least temporary relief: Alaska in 1975 and fracking in 2008. For how long can we count on such miracles? Will the world shift to non-carbon sources of energy if relatively cheap oil is still available? The next quarter to half century is likely to be extremely interesting with respect to energy, although I have hardly a clue to predict how. My guess is that whatever unfolds with respect to energy limitations will be regional, not global, and that social unrest will be the main way that such restrictions are translated into impact on society, as has been beautifully analyzed by Ahmed (2017) in another book in this series.

12.4 EROI

Of perhaps greater concern than the quantity of oil and other energy sources is their declining EROI. The world will not run out of hydrocarbons (e.g., Lynch 2002). Instead it has, and will increasingly, become difficult to obtain cheap petroleum, because what is left is an enormous amount of low-grade hydrocarbons which are likely to be much more expensive financially, energetically, politically, and environmentally. As conventional oil becomes less available, society probably will make investments in different sources of energy and improvements in energy use efficiency, in theory reducing our dependence on hydrocarbons (Kaufmann 2004; Swensen 2016). There is already a substantial decrease in the energy we use per unit

12.4 EROI

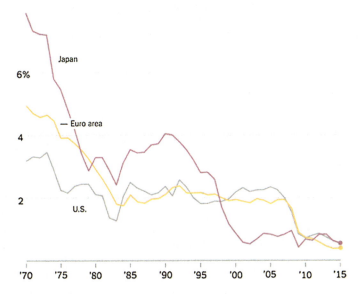

Fig. 12.6 GDP growth rates of selected OECD entities. Each point reflects average growth over preceding 10 years. *Source* World Bank. Compare with Fig. 9.4

of GDP in the United States, but Kaufmann found that most of the reason was because we had switched much of our economic "heavy lifting" (heavy industry) overseas, not for any intrinsic increase in efficiency. The energy efficiency of the world has not changed too much (Fig. 12.10b). Those who focus on sustainability and protecting our atmosphere and climate cheer this process on with great enthusiasm. But one needs to be a little careful about what one wishes for. It is extremely unlikely that we will ever have another major energy resource as valuable, abundant, and flexible as oil and gas. The rate of increase in the use of these fuels, previously some 3–5 % per year, has declined precipitously, especially in Western economies, to less than 1 % per year even as prices increased then declined (Fig. 12.3; also Fig. 9.4). The reasons for this are several, including increased efficiencies, but appear to be mostly due to the fact that the economies of Europe, Japan and to a lesser degree the US and China experienced much slower rates of economic growth (Fig. 12.6). This may be due to the relatively high price of oil (especially outside the United States). While the price of oil seems low now it remains high by historical standards. How this will affect future use rates of fossil fuels remains uncertain.

As the energy required to find and deliver high-quality energy becomes larger at the societal level, there may be too little energy surplus available for other activities or insufficient usable energy to drive economic growth (Cleveland et al.1984; Ayres and Voudouris 2014). In the literature, there is widespread concern that net energy returns (e.g. EROI) for oil and gas are declining and likely to continue declining (e.g., Hall and Cleveland 1981; Dale et al. 2011; Grandell et al. 2011; Brandt et al. 2013). Some analysts also raise the issue of a low net energy return from the rapid scale-up of low-carbon energy technologies (Arvesen et al. 2011; Carbajalis-Dale and Benson 2013). Meanwhile the energy costs for adjusting to degrading environments escalates (e.g. Vitousek et al. 2001; Day and Hall 2016).

12.5 What Level EROI Does Society Need?

As with our cheetah, society does not need a "just barely sufficient" EROI to survive but rather one with a significant net profit. Quoting with slight modification from Hall (2011): "Think of a society dependent upon one energy resource: its domestic oil. If the EROI for this oil was 1.1:1 then one could pump the oil out of the ground and look at it. If it were 1.2:1 you could also refine it and look at it, 1.3:1 also distribute it to where you want to use it but all you could do is look at it. Hall et al. (2014) examined the EROI required to actually run a truck and found that if the energy included was enough to build and maintain the truck and the roads and bridges required to use it (i.e., depreciation), one would need at least a 3:1 EROI at the wellhead to put one unit of gasoline into the truck. Now if you wanted to put something in the truck, say some grain, and deliver it that would require an EROI of, perhaps, 5:1 to grow the grain. If you wanted to include depreciation on the oil field worker, the refinery worker, the truck driver and the farmer you would need an EROI of say 7 or 8:1 to support the families. If the children were to be educated you would need perhaps 9 or 10:1, have health care 12:1, have arts in their life maybe 14:1 and so on (the numbers below 3:1 are fairly accurate, and above are speculative). Obviously to have a modern civilization one needs not simply surplus energy but lots of it, and that requires either a high EROI or a massive source of moderate EROI fuels" (Fig. 12.7). Thus, as the EROI of alcohol derived from corn was from 1:1 to 1.6:1 (Chap. 11), it is very unlikely one could support much of a complex civilization. If the entire energy "food chain" required to support the farmer, distiller, and so on were included, there would not be much in the way of a net yield to the rest of society. Lambert and colleagues estimated that the EROI to run modern industrial-consumer societies is probably much higher, probably from 10:1 to 15:1 at a minimum if we are to support families, health care, education, the more complex arts, and so on. Unfortunately, most liquid alternatives to oil available today, or perhaps wind and solar PV if appropriate corrections are made for their intermittency, are characterized by EROIs much lower than this, limiting their economic value (Lambert et al. 2014; Keifer 2013). Consequently, it remains a matter of speculation whether renewable energy sources, even with the rapid decline in their costs, could sustain high levels of economic output and continued growth (Kaufmann and Shiers 2008). It is critical for CEOs, government officials and the public to understand that the best oil and gas are simply gone, and there is no easy replacement. We may be able to adjust to this situation, but not if we think

12.5 What Level EROI Does Society Need?

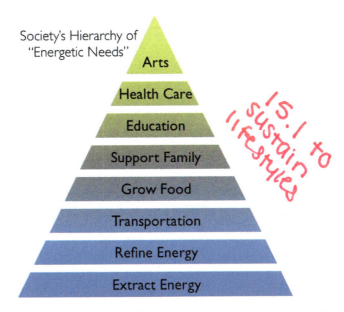

Fig. 12.7 "EROI pyramid" of increasing abilities to support economic activities as a function of the EROI of the main fuels for the economy. The values run from 1.1:1 to extract energy, 3:1 to provide transportation, etc. to perhaps 12 or 15:1 to provide complex arts, as defined in the text. The values up to transportation are measured, and those above are increasingly speculative. Graph from Lambert et al. 2014 as inspired by Maslow 1943

that high levels of, or perhaps any, economic growth are necessarily going to happen. These are powerful arguments to be considered when thinking about sustainability.

12.6 Economic Impacts of Peak Oil and Decreasing EROI

It is well known that there is a strong correlation between a nation's, and the world's, energy use and material well being (Fig. 12.8). There also is a strong correlation between a nation's total energy use, its EROI and the equity of its distribution, and both GDP and more importantly various indices of human well-being, such as the HDI (human development index), at least up to some point (Figs. 12.9 and 12.10). If we are to maintain or increase human well-being into a future of possible decreasing energy availability, we will have to think very carefully about how to do so. A more equitable distribution of energy resources is one logical place to start.

Whether global peak oil has occurred already, or will not occur for some years or, conceivably, decades, its economic implications will be enormous (Kopits 2014). Hamilton (2009) has found that all economic recessions in recent decades in

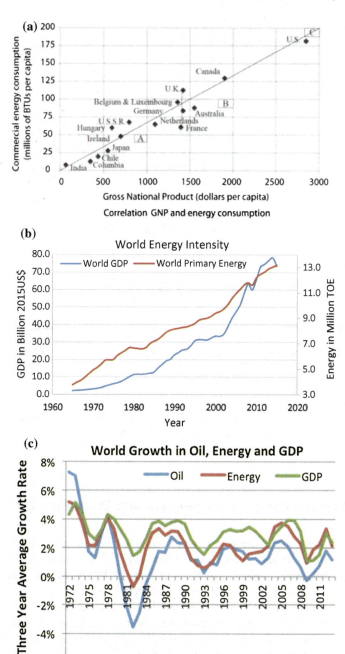

Fig. 12.8 a The relation of the GDP to energy use for selected countries. **b** The growth of the world economy from 1965 to 2015 has been enormous, as has been the use of energy (*Source* G. Tverberg). **c** Energy and GDP growth rates of the world, showing high correlation and general decline of all. (*Source* Gail Tverberg)

12.6 Economic Impacts of Peak Oil and Decreasing EROI

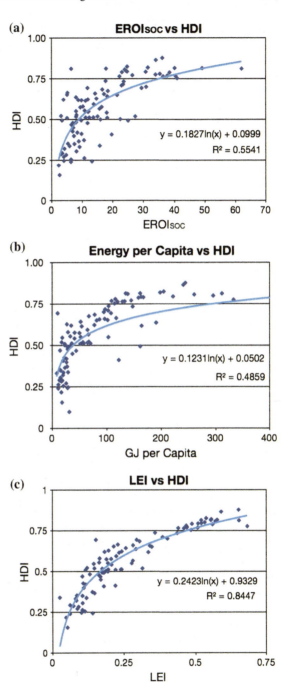

Fig. 12.9 Correlation among nations of human development index (HDI – designed to be an alternative to GDP for measuring human well being) with (a) EROI for that nation, (b) energy use per capita and (c) Lambert Energy Index, a composite energy index which includes per capita use, EROI and equitability of use

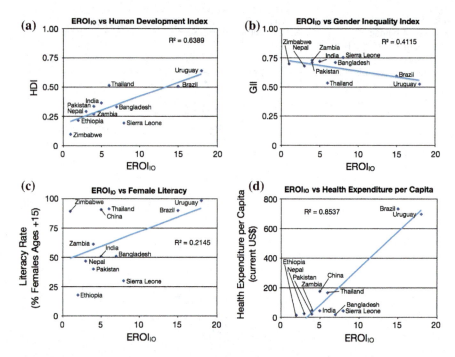

Fig. 12.10 Human well-being (as imperfectly measured by the human development Index) versus several indices of energy availability. These are combined in the "Lambert energy index." which includes per capita energy, EROI of that energy and the equitability of its distribution (from Lambert et al. 2014)

the United States had been preceded by an increase in the price of oil. As oil becomes less readily available in future decades, we do not know if there are substitutes available on the scale required and at the EROI that is needed. Alternatives will require enormous investments in money and energy, both likely to be in short supply. Despite the projected impact on our economic and business life within relatively few years in the USA, neither government nor the business community is prepared to deal with either the impacts of these changes or the new thinking needed for investment strategies (Hall and Klitgaard 2012; Hirsch et al. 2005; Hall and Groat 2010). The reasons are myriad but include: the vested interests of powerful energy companies, the disinterest and disorientation of the media, the erosion of good energy record keeping at the Department of Commerce, and the focus of the media on trivial "silver bullets" despite the inability of any one of them (except economic contraction and in some cases conservation) to contribute more than a few percent to the total energy mix, and the failure of government to fund good objective analytic work on the various energy options. Consequently much of what is written about energy is woefully misinformed or simply advocacy by various groups that hope to look good or profit from various perceived alternatives. The end of cheap petroleum will be perhaps the most important challenge that Western society has ever faced, especially when considered within the context of the need to deal with climate change, growing populations, and aspirations and other environmental issues related to energy (Jones and Warner 2016). Business and government leaders who do not understand the inevitability, seriousness, and implications of the end of cheap oil and eventually cheap fossil fuels, or who make poor decisions in an attempt to alleviate their impact, are likely to be tremendously and negatively affected as a result. At the same time the investment decisions made in the next decades will determine whether or not civilization is to make it through the transition away from petroleum driven, growth-based economies to something more sustainable, if indeed that is possible.

12.6.1 Secular Stagnation

From 1965 to 2015 the world economy, and its energy use, increased substantially (Fig. 12.10) even while some countries appeared to be getting more efficient (Fig. 12.11). But the growth of most of the world's industrialized economies have declined enormously in recent years (Figs. 12.6 and 9.4). As of mid-2016 the GDP of countries in Europe and Japan had been essentially stagnant for a decade or two. The United States had a GDP growth rate of 1.1 %, extremely low by historical standards and about the same as the population growth—hence no average increase in per capita wealth. Amongst economists there is considerable discussion and controversy about this (e.g., Galbraith 2014; Irwin 2016). Much of this focuses on factors internal to the economy: consumer spending, debt, banks, deficit spending and Keynesianism—whether or why governmental deficit spending, which has been used extensively in the past to "jump start" economic growth, no longer works as it once did. There is nothing consistent in conventional economics that has an explanation for this general secular stagnation. It is possible that a new approach to economics called BioPhysical Economics, which attempts to make economics based on the natural sciences rather than the social science-based discipline that it is, may provide such explanation (Hall and Klitgaard 2006; Hall and Klitgaard 2012). Most adherents to biophysical economics believe (as do many others) that conventional (neoclassical) economics is fundamentally flawed. According to Hall et al. (2001) conventional economics cannot possibly be accepted by people trained in the natural sciences because:

(1) its basic model violates the laws of thermodynamics
(2) the boundaries used for analysis are incorrect and
(3) its basic premises are put forth as givens rather than tested hypotheses.

BioPhysical economics believes that there is a general relation between the declining abundance of resources, as reflected in lower production and EROI for oil and other important fuels, and the decline and cessation of growth (Figs. 9.4 and 12.6). Murphy and Hall (2011) put forth a model that gave a biophysical economics-based explanation of economic cycles that seems consistent with the

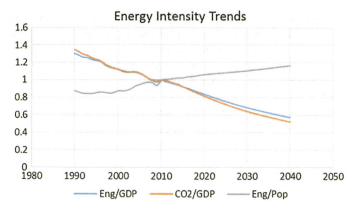

Fig. 12.11 Energy efficiency, derived as energy used per unit GDP, has decreased in recent decades, even as energy use has greatly increased. (Source Gail Tverberg)

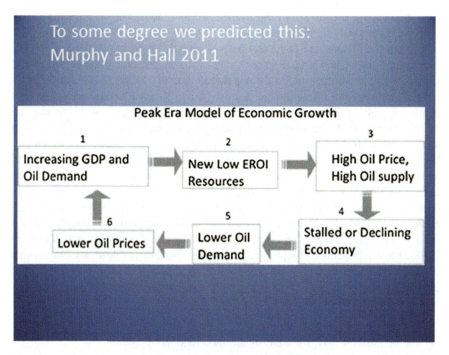

Fig. 12.12 Simple biophysical model of the relation between EROI and economic activity. (*1*) During a period of relatively large economic growth (e.g.,1990s and early 2000s) the increasing demand for oil (*1*) to meet the requirement for more economic activity, requires the exploitation of more, lower EROI, resources which are more expensive (*2*). This leads to higher oil price (*3*, as in 2014) and a stalled economy (*4*) which leads to a lower oil demand (*5*) and lower oil prices (*6*) which eventually leads to more economic growth (*1*)

12.6 Economic Impacts of Peak Oil and Decreasing EROI

actual behavior of economies (Fig. 12.12). The case for this was stronger up to mid-2015, when oil was trading at $100 a barrel. At the time of this writing it is about $50 a barrel, still high by historical standards and relative to, e.g., the 1990s when growth was still strong. The OECD country with highest growth, although still low, is the United States. In the US natural gas, not quite as valuable as oil but still an excellent fuel for industry, was at a very low price, about a quarter of the long-term price, reflecting over production from fracked areas in, e.g., the Marcellus Shale in Pennsylvania. This could be the reason for the slightly higher growth of the U.S. economy compared to other OECD countries.

There may be another useful biophysical concept from ecology. Eugene Odum in 1969 wrote a good paper representing the behavior of ecosystems over successional time, that is from the establishment or colonization of life at a particular site, such as an ecosystem that develops on a bare patch of land or an empty aquarium until the ecosystem reached "climax," when it no longer accumulated biomass. At first, as biomass became established, production (the capture of energy from sunlight) and respiration (the use of energy for maintenance metabolism by all living things) each increased rapidly, with photosynthesis being larger than respiration (Fig. 12.13). The difference between the two represented the energy absorbed by the increasing biomass. But then at some point the respiration of the increasing biomass equaled the production of the plants and the system stopped accumulating biomass. The relatively constant biomass remaining at steady state was limited by

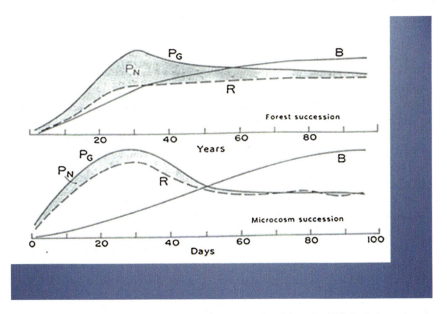

Fig. 12.13 Ecological concept of succession following disturbance (or initiation); at some point increased infrastructure (biomass) requires all energy available for maintenance metabolism and growth ceases. *Top graph* for a forest starting on bare ground; *bottom graph* for an aquarium with many species

the incoming solar energy, and the system adapted to that. This takes place around the world as most ecosystems are limited by the incoming solar radiation (or water) and rather than growing indefinitely reach a steady state biomass level. Odum believed that human societies too would initially grow rapidly ($P \gg R$) but then would approach equilibrium as energy costs to maintain infrastructure became very large. This is very different from the indefinite growth of economies expected by most economists. So, have modern highly developed economies with enormous infrastructures (think roads and cities) reached a stage where all of the available energy is used in "maintenance metabolism" to support the infrastructure that exists and little is available for net growth? Could this be an explanation for secular stagnation? Or is it sufficient to say that the growth of economies simply reflects the growth of the energy that allows that to happen, and that energy, once easy and cheap to get, is no longer so? Either way Bio Physical economics has approaches that appear very useful to understanding secular stagnation which should be explored much more than they have been so far.

12.7 Developing Energy Policy

Developing comprehensive energy policy is highly complex and is beyond the scope of this chapter. However, there is a set of questions arising from certain characteristics of energy developed in this book that should be asked when any energy source is proposed or whenever the word sustainable is used.

1. Is the proposed plan for exploiting a new energy source consistent with the laws of thermodynamics?
2. Are external energy subsidies, including those from environmental degradation, accounted for?
3. Has a proper EROI been calculated and is the EROI 10:1 or better as required for a viable modern economy?
4. Is the produced energy easily available to energy consumers, or is there a mechanism to transfer energy from the site of production to site of consumption and how expensive, monetarily and energetically, is it?
5. Is the produced energy available when needed, or does a storage option exist (for wind and solar produced electricity, for example)? Is the output synchronized with consumption patterns?
6. Is the energy produced in a form that is easily used, stored and transported, and what are the main safety issues involved? Are the costs of storage etc, included in the EROI?
7. Do the risks of producing and handling the energy exceed the potential benefits?
8. Is there a key substance in the production process that is the limiting factor? For example the electric utility industry has concerns that copper supplies are insufficient to produce the transmission lines needed to support a transportation system reliant on electricity.

12.7 Developing Energy Policy

9. And more generally does the plan (and whatever other sources are operating) allow for adequate net energy to run the desired economy? Are there sufficient funds or available debt and/or surplus energy to in fact allow it to be implemented? Will the output be worth the investment?

It will be necessary for a successful transition from the petroleum age that sufficient alternative energy sources be found for which all these questions can be answered in the affirmative. If not, what then?

Whatever the future will be it is likely to be one of very large change. Climate is likely to be very different, in many cases imposing very large requirements for more energy to adapt to it. There will be many pressures to replace fossil fuels with solar-based technology, especially solar PV and wind, which would require enormous amounts of capital expenditures for the transition if that is possible. Heinberg and Fridley (2016) think it is inevitable, even though they calculate that it would take a rate of capital investments in new technologies 20 times our present rate. Capital expenditures mean energy investments. Given that total energy use may be plateauing (Fig. 9.5) this would mean an enormous diversion of energy for these capital expenditures, both for adapting to climate change as well as for alternative energy infrastructures. McKay (2010) also gives a very credible assessment of the possibilities and difficulties of moving to a renewable society.

Meanwhile, the main problem that we face with regard to fossil (and other) fuel supplies is not their total quantity on Earth (there are enormous supplies remaining) but their quality. To survive and thrive, all species must balance the relation between the energy cost of getting needed resources, including additional energy, and the energy (or other attributes) in the resource exploited. This applies to predators hunting for food who must compare the energy expended in the chase and the chances of success with the energy obtained from the prey. Likewise human hunter gatherer societies, if they are to survive, must generate substantial surpluses of energy relative to their own investment energies. It also applies to modern human industrial economies, although they are different in that the energy invested and gained is not metabolic but exosomatic (outside the body) energy. Thus, a critically important issue for examining our energy future is what is called energy return on investment (EROI) of fuels. Investments are required to get fossil and other fuels out of the ground and into society. These investments are in terms of energy as well as dollars and just as we need a profit from a financial investment so we need a net energy profit from our energy resources for society to continue.

The most important question for us is how we should make our energy investments. Huge investments will be needed if we are to maintain the enormous human infrastructure that we have built simply to fight the inevitable generation of entropy which nature dictates will occur. As we move into the future EROI is a critically important component of the decisions we have to make, but hardly sufficient by itself. The main problems that we face at this time with respect to understanding our situation are as follows:

(1) The apparently incessant decline in EROI which will greatly limit our options for investing in new energy technologies, whatever they might be (Fig. 12.14).

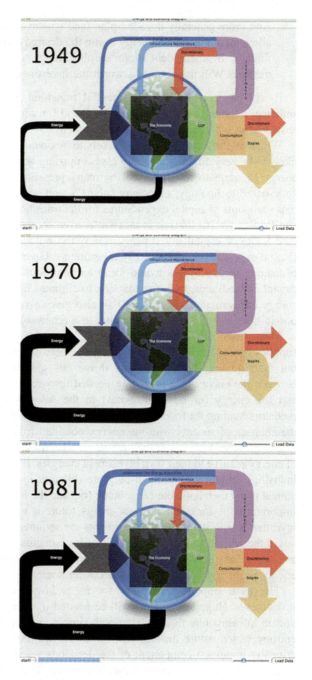

Fig. 12.14 Past proportional disposition of U.S. national GDP (in dollars, with the total output of the economy represented (in green) as 100 %) and potential decline in usable energy, and the possibilities for investments in growth, as EROI declines into the future (from Hall et al. 2008). Note that as requirements for more and more energy increases as EROI declines (dark blue: e.g. as during the "energy crisis" of 1970-1980)), the amount left for maintenance metabolism (light blue) and other investments (red) declines. 'Staples' (yellow) means minimal official requirements for food, shelter and clothing.

12.7 Developing Energy Policy

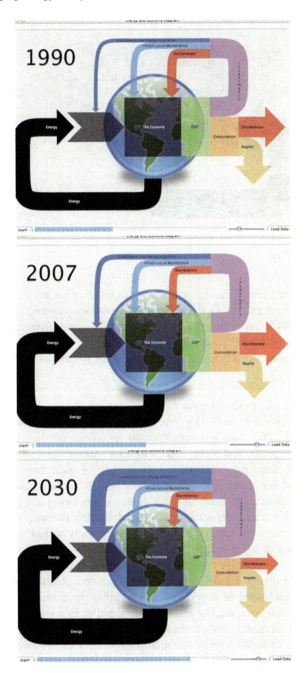

Fig. 12.14 (continued)

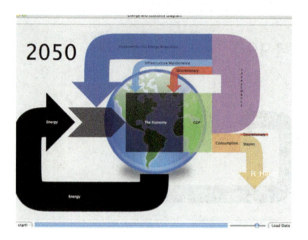

Fig. 12.14 (continued)

(2) The need for some kind of professional, objective means of gathering the needed data and evaluating the alternative energy sources and claims. It would seem that such an evaluation would have to come from peer reviewed, government-sponsored program.
(3) The total inadequacy of conventional economics for the job (Fig. 12.15).

Summary

Our expectations for our lives for the past several hundred years have been based on an expanding universe of lands (e.g., the Americas), energy, and energy returns on energy invested. This has generated in the psyches' of the minds of most Americans, Europeans and East and Southeast Asians, and many more, the expectation that there was at least the possibility of their bettering their own material lot, and for many this occurred. We hear it in the pronouncements of economists and politicians from all sides, how we are facing a tragedy as young people no longer have an expectation of more than their parents. To some degree this is clearly because the reins of power have increasingly passed into the hands of the wealthy. Most of us no longer live in a democracy but rather a plutocracy. But something else is happening too: Malthus is finally catching up with us, if not exactly now (it probably is) then it is likely to come on in spades soon. The global population can no longer be supported without piling up enormous debt, in monetary terms but also energetically and environmentally. Everywhere we look there are serious environmental issues starting with the potential impacts of climate change. There is certainly a lot of attention paid in some circles to climate change. But I believe the potential impact on our future society from issues related to energy supply and EROI, however, they work out, are likely to be as large or even larger. Hopefully the understanding and use of EROI in analyses and public media will help soften the hard landing ahead of us, as the fossil fuels that have allowed many of the world's 7.3 billion people to live in relative luxury by the standards of old

12.7 Developing Energy Policy

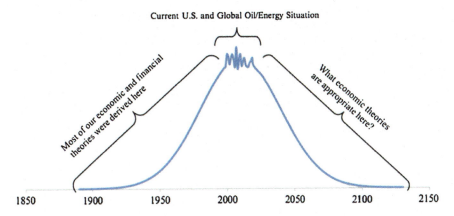

Fig. 12.15 Diagram of relation of the development of economic and financial theories to energy conditions

inevitable deplete. But neither our economists nor our politicians have the conceptual base or mental models to deal with this, and still rely on mental models where the only operational levers for society are within the economy. Rather economists must understand that much of what has determined human history, and is likely to continue to do so, comes from outside the immediate economy and is far less susceptible to internal manipulation (Fig. 12.14). BioPhysical economics is one antidote to this, but hardly sufficient. We need an entirely new approach to education, including how we can work together to face a world with increased constraints on our energy and economic growth (Fig. 12.15).

References

Adelman, M.A., and M.C. Lynch. 1997. Fixed view of resource limits creates undue pessimism. *Oil & Gas Journal* 95: 56–60.

Ahmed, N. 2017. *Failing states: BioPhysical triggers of political violence*. NY: Springer Briefs.

Arvesen, A., R.M., Bright, and Hertwich, E.G. 2011. Considering only first-order effects? How simplifications lead to unrealistic technology optimism in climate change mitigation. Energy Policy 39: 7448–7454.

Ayres, R., and V. Voudouris. 2014. The economic growth enigma: capital, labour and useful energy? *Energy Policy* 64: 16–28.

Brandt, A.R. 2007. Testing Hubbert. *Energy Policy* 35: 3074–3088.

Brandt, A.R., M. Carbajalis-Dale, and C. Barnhart. 2013. Calculating systems-scale energy efficiency and energy returns: a bottom-up matrix-based approach. *Energy* 62: 235–247.

Campbell, C.J., and J.H. Laherrère. 1998. The end of cheap oil. *Scientific American* 278: 78–83.

Carbajales-Dale M., M. Raugei, C.J. Barnhart, and V. Fthenakis. 2015 Energy return on investment (EROI) of solar PV: an attempt at reconciliation. Proceedings of the IEEE. http://dx.doi.org/10.1109/JPROC.2015.2438471.

Carbajales-Dale, M., and S.M. Benson. 2013. The energy balance of the photovoltaic (PV) industry—Is the PV industry a net energy provider? *Environmental Science and Technology* 47(7): 3482–3489.

Cleveland, C.J., R. Costanza, C.A.S. Hall, and R. Kaufmann. 1984. Energy and the United States economy: a biophysical perspective. *Science* 225: 890–897.

Dale, M., S. Krumdieck, and P. Bodger. 2011. Net energy yield from production of conventional oil. *Energy Policy* 39(11): 7095–7102.

Day, J. W., and C.A.S. Hall. 2016. America's most sustainable cities and regions: Surviving the 21st century megatrends. Springer, N.Y.

Diamond, J. 2005. *Collapse*. Collapse: How Societies Choose to Fail or Succeed. Viking N.Y.

Durden, T. 2016. Debt, Deficits & Economic Warnings, Aug 30, 2016 6:10 PM. http://www.zerohedge.com/news/2016-08-30/debt-deficits-economic-warnings.

Irwin, N. 2016. We're in a Low-Growth World. How Did We Get Here? The Upshot Aug. 6, 2016.

Galbraith, J. 2014. The end of normal: The great crisis and the future of growth. Simon and Shuster N.Y.

Goodland, R., and H. Daly. 1996. Environmental sustainability: universal and non-negotiable. *Ecological Applications* 8: 1002–1017.

Grandell, L., C.A.S., Hall, and M. Hook. 2011. *Energy return on investment for Norwegian oil and gas in 1991–2008*. Sustainability: Special Issue on EROI, pp 2050–2070.

Gagnon, N., C.A.S. Hall, and L. Brinker. 2009. A preliminary investigation of energy return on energy investment for global oil and gas production. *Energies* 2(3): 490–50.

Hall, C.A.S. 2011. Introduction to special issue: Sustainability: Special Issue on EROI. pp 1773–1777.

Hall, C. 2016. Review of Heinberg and Fridley. BioScience December 2016.

Hall, C.A.S., S. Balogh, and D.J.R. Murphy. 2009. What is the minimum EROI that a sustainable society must have? *Energies* 2: 25–47.

Hall, C., and K. Klitgaard. 2006. The need for a new, biophysical-based paradigm in economics for the second half of the age of oil. *Journal of Transdisciplinary Research* 1(1): 4–22.

Hall, C., and K. Klitgaard. 2012. *Energy and the wealth of nations*. Springer, N.Y.: Understanding the Biophysical Economy.

Hall, C., P. Tharakan, J. Hallock, C. Cleveland, and M. Jefferson. 2003. Hydrocarbons and the evolution of human culture. *Nature* 426: 318–322.

Hall, C.A.S., and C.J. Cleveland. 1981. Petroleum drilling and production in the United States: Yield per effort and net energy analysis. *Science* 211: 576–579.

Hall, C.A.S., R. Powers and W. Schoenberg. 2008. Peak oil, EROI, investments and the economy in an uncertain future. Pp. 113-136 in Pimentel, David. (ed). Renewable Energy Systems: Environmental and Energetic Issues. Elsevier London.

Hall, C.A.S., and A. Groat. 2010. Energy price increases and the 2008 financial crash: a practice run for what's to come? *The Corporate Examiner*. 37(4–5): 19–26.

Hall, C.A.S., D. Lindenberger, R. Kummel, T. Kroeger, and W. Eichhorn. 2001. The need to reintegrate the natural sciences with economics. *BioScience* 51: 663–673.

Hall, C.A.S., J.G., Lambert, and S.B. Balogh. 2014. EROI of different fuels and the implications for society. *Energy Policy Energy Policy* 64: 141–152.

Hamilton, J.D. 2009. *Causes and consequences of the oil shock of 2007–08*. Brookings Papers on Economic Activity, Spring.

Hallock, J., P. Tharkan, C. Hall, M. Jefferson, and Wu, W. 2004. Forecasting the limits to the availability and diversity of global conventional oil supplies. *Energy* 29: 1673–1696.

Hallock, Jr, L. J., W. Wei, C. A.S. Hall, and M. Jefferson. 2014. Forecasting the limits to the availability and diversity of global conventional oil supply: Validation. *Energy* 64: 130–153.

Heinberg, R., and D. Fridley. 2016. *Our renewable future: Laying the path for one hundred percent clean energy*. Washington, D.C.: Island Press.

Hirsch, R., R. Bezdec, and Wending, R. 2005. *Peaking of world oil production: impacts, mitigation and risk management*. U.S. Department of Energy. National Energy Technology Laboratory. Unpublished Report.

Holter, M. 2016. *Oil discoveries at 70-year low signal supply shortfall ahead*. Bloomberg The year ahead. August 30.

References

Irwin, N. 2016. *We're in a low-growth world. how did we get here?* The upshot Aug. 6, 2016.

Jacobson, M.Z., and M.A. Delucchi, November 2009. *A path to sustainable energy by 2030*, Scientific American.

Jacobson, M.Z., M.A. Delucchi, G. Bazouin, Z.A.F. Bauer, C.C. Heavey, E. Fisher, S.B Morris, D.J.Y. Piekutowski, T.A. Vencill, T.W. Yeskoo. 2015. 100 % clean and renewable wind, water, sunlight (WWS) all-sector energy roadmaps for the 50 United States. *Energy and Environmental Sciences* 8: 2093–2117.

Jones, G.A., and K.J. Warner. 2016. The 21st century population-energy-climate nexus. *Energy Policy* 93: 206–212.

Kiefer, T.A. 2013. *Energy insecurity: The false promise of liquid biofuels*. Montgomery: Air Force Research Institute.

Kaufmann, R.K., and L.D. Shiers. 2008. Alternatives to conventional crude oil: When, how quickly, and market driven? *Ecological Economics* 67: 405–411.

Kaufmann, R.K. 2004. The mechanisms for autonomous energy efficiency increases: A cointegration analysis of the US energy/GDP ratio. *The Energy Journal* 25(1): 63–86.

Kaufmann, R.K., and C.J. Cleveland. 2001. Oil Production in the lower 48 states: economic, geological and institutional determinants. *Energy J.* 22: 27–49.

Kopits, S. 2014 Lecture at Columbia University: https://www.youtube.com/watch?v=dLCsMRr7hAg.

Laherrere, J. 2016. ASPO France web page.

Lambert, J., C.A.S. Hall, S. Balogh, A, Gupta, and M. Arnold. 2014. Energy, EROI and quality of life. *Energy Policy Volume* 64: 153–167.

Lynch, M.C. 2002. Forecasting oil supply: Theory and practice. *The Quarterly Review of Economics and Finance* 42: 373–389.

McGlade, C., and P. Ekins. 2015. The geographical distribution of fossil fuels unused when limiting global warming to 2 degrees C. *Nature*.

Maggio, G., and G. Cacciola. 2012. When will oil, natural gas, and coal peak? *Fuel* 98: 111–123.

Maslow, A. 1943. a theory of human motivation.Psychological Review 50: 370–396.

McKay, D. 2010. *Renewable energy—Without the hot air*. London: UIT Press.

Mohr, S.H., J. Wang, G. Ellem, J. Ward, and D. Giurco. 2015. Projection of world fossil fuels by country. *Fuel* 141: 120–135.

Murphy, D.J., and C.A.S. Hall. 2011. Energy return on investment, peak oil, and the end of economic growth. Annals of the New York Academy of Sciences. *Special Issue on Ecological economics* 1219: 52–72.

Nashawi, I.S., A. Malallah, and M. Al-Bisharah. 2010. Forecasting world crude oil production using multicyclic Hubbert model. *Energy & Fuels* 24: 1788–1800.

Odum, E.P. 1969. The strategy of ecosystem development. *Science* 164: 262–270.

Simmons, Matt (No date) *The world's giant oil fields*. Simmons and Company International.

Smil, Vaclav. 2011. Global energy: The latest infatuations. *American Scientist* 99: 212–219.

Swensen, R. 2016. The solar evolution: Much more with way less, right now—The disruptive shift to renewables. *Energies* 2016(9): 676.

Tainter, J. 1988. *The Collapse of Complex Societies*. Cambridge: Cambridge Univ. Press.

Trainer, T. 2013. Can Europe run on renewable energy? A negative case. *Energy Policy* 63: 845–850.

Tripath, V. S. and A. R. Brandt. (In Revision). Estimating Decades-Long Trends in Petroleum Field Energy Return on Investment (EROI) with an Engineering-Based Model. PLOS ONE.

United States Geological Survey (USGS). (2003). *The World Petroleum Assessment 2000* www.usgs.gov.

Vitousek, P.M., H.A. Mooney, J. Lubchenco, and J.M. Melillo. 1997. Human domination of Earth's ecosystems. *Science* 277(5325): 494–499.

Index

A
Agriculture, 91, 92, 129
 early, 92, 93
 energy cost, 93

B
Biomass energy, 18, 81, 82, 122, 151, 161
 energy source for early civilizations, 60
Biophysical, basic concept, 146, 161
Biophysical economics, 101, 102, 159, 167
 definition, 102, 159
 need for, 101, 102

C
Campbell
 Colin, 41, 146–148
Capital, 124, 163
 carbon
 cost, 85, 124
 energy cost, 124
Carnot, Sadi, 10, 11, 13, 14
Climate, 35, 42, 63, 70, 85, 91, 132, 146, 149, 151, 153, 159, 163, 166
Coal, 7–10, 22, 23, 26–28, 39, 41, 42, 44, 45, 75, 90, 93, 95, 97, 100, 110, 114, 115, 119, 122–125, 127, 131, 134, 136, 137
Colonialism, 5, 8, 84, 161
Competition, 65, 73, 74, 85, 101
Copper, 53, 109, 110, 135, 138, 162
Corn (Maize), 27, 37, 108, 128, 129, 131, 132, 154

D
Dams, 22
 EROI
Deforestation, 91
 Perlin, 92
Deindustrialization

Depreciation, 71, 124, 127, 154
 and energy cost, 71, 124, 127
Dissipative structures, 66

E
Ecology, 50, 62, 63, 85, 161
 and evolution, 48–50, 65, 66
Economics, 3–5, 8, 16, 22, 27, 29, 45, 66, 69, 75, 83–85, 90, 92, 95–102, 107–110, 114, 116, 124, 129, 132, 145, 150, 153–155, 158, 159, 162, 166, 165
 biophysical, 101, 102, 159
 critiques, 119, 129, 138
 growth, 22, 29, 45, 101, 102, 145, 146, 153–156, 159, 160, 162, 167
 neoclassical, 85, 100, 101, 159
Ecosystems, 18, 22, 38, 39, 54, 67, 74, 75, 81, 82, 90, 137, 161, 162
 industrial, 100, 107, 124, 159, 163
 succession, 161
Efficiency, 9–11, 19, 27, 57, 73–77, 81–83, 85, 97, 99–101, 109, 110, 114, 134, 139, 152, 153, 160
 maximum, 11, 73, 76
 of steam engines, 10, 11, 137
 thermodynamic, 9–11, 19
Einstein, Albert, 15
Electric power plant, 12, 15, 29, 96, 134
 efficiency, 18, 19, 24
Embodied energy, 28, 128
Energy and
 agriculture, 5, 90, 91
 history, 7, 35, 54, 89, 90, 92
 mining, 93, 109
Energy budget, 121
 fish, 67
 stream, 19
Energy carriers, 21

Energy cost, 18, 49, 55, 65, 67, 68, 70, 78, 92, 93, 101, 110, 111, 122–129, 132, 133, 135–137, 139, 140, 154, 162, 163
   energy, 27, 30, 65, 67, 92, 122–125, 129, 132, 133, 136, 137, 163
   evolution, 49, 50, 54, 65
   infrastructure, 92, 162
   society, 92, 93, 110, 121, 124, 127, 129, 163
Energy definition, 23
Energy/GNP ratio, 124
   used to derive indirect energy costs, 124
Energy gradient, 23, 28, 52, 72
Energy opportunity cost, 4, 72
Energy prices, 110
   and EROI, 109
Energy quality, 27, 119, 120, 133, 134
   electricity, 27, 119, 120, 133, 134
Energy return on investment (EROI), 5, 43, 66, 67, 70–72, 91–93, 107–116, 125, 127–140, 146, 152, 154, 155, 158, 159, 163, 166
   calculation, 127, 130
   critiques, 119, 129, 138
   coal, 93, 119, 125, 131, 134, 137
   definition, 66, 71, 133, 163
   energy quality, 119–121, 133, 134
   food capture, 127
   hunter-gatherer, 90
   insulation, 125
   nuclear power, 134
   of different fuels, 111, 115, 119, 127
   oil, 43
   oil shale, 43
   renewable, 137
   solar, 91, 108, 131, 135
   wood plantations
Energy storage, 28, 136, 137
   batteries, 28
   difficulties, 137
   pumped storage, 137
Energy Type, 28
   Kinetic, 28
   Potential, 28
   Embodied, 28
   Primary energy source, 28
   Carrier, 28
Entropy, 12–16, 18, 26, 52, 54, 66, 89, 163
   and heat, 12–16, 18, 26, 66
   and second law of thermodynamics, 14, 15
   definition, 12
Environment, 15–18, 26, 39, 41, 49, 50, 53–55, 61–67, 71, 72, 74, 77, 79, 81, 82, 90–92, 97, 102, 125, 136, 138, 145, 152, 154, 159, 166
Evolution, 5, 22, 29, 38, 49, 50, 54, 55, 59–62, 65–67, 70, 71, 73–75, 84, 89–91, 96, 107
   and energy, 5, 22, 28, 38, 49, 50, 54, 55, 59, 62, 65–67, 70, 71, 73–75, 107
   and EROI, 5, 66, 70, 71, 73, 91, 107
   cultural, 60, 90, 91
   organic, 55, 74, 90, 91
Externalities, 17, 129, 130

F
Falkowski, Paul, 55
Fertilizer, 30, 42, 132

G
Gross domestic product (GDP), 4, 93, 98, 99, 139, 153, 155–157, 159, 160, 164
Growth
   economic, 22, 29, 45, 101, 102, 145, 153, 155, 159, 160, 167
   of organisms, 5, 18, 55, 66, 67

H
Hadley cells, 36, 38
Homeostasis, 55
Hubbert curve
   validation, 146–150
Hubbert, Marion King, 146
Hubbert production cycle, 146–150

I
Imports
   EROI of, 129
Industrialization
   history, 45
Inflation, 100–110, 124, 128

J
Joule
   as preferred means of measuring energy, 14, 25, 26, 29, 31, 111
Joule, James, 13, 14

K
Keynes, John Maynard
Kung, 90, 91
   EROI, 91

Index

L
Labor, 92, 93, 100, 126, 139, 140
Laherrere, Jean, 122, 146, 149
Lavoisier, Antoine, 10
Liebig law of minimum
Limits to growth, 18, 22
Liquefied natural gas, 42, 122, 123, 147, 148
Louisiana, 41, 42

M
Maintenance metabolism or respiration, 55, 56, 161
Markets, 99, 101
Mathematics, 8
Maximum power, 73–78, 80, 81, 83–86
Maxwell, James Clerk
Model, 85, 100, 146, 148, 150, 159, 160, 167

N
National debt, 129, 159, 163, 166
Natural energies, 18, 22, 26, 28, 90, 111
Natural gas
  liquids, 25, 42, 122, 123, 147, 148, 150
Natural resources, 109, 110
Natural selection, 60
  natural selection and EROI, 5, 66, 67, 69, 72, 73, 86, 91–93, 104, 107–116, 125, 127–140, 145, 146, 150, 152, 154, 155, 157–164, 167
Nature, 8, 10, 13, 23, 28, 38, 41, 49, 59, 62, 64, 65, 69, 70, 74, 77, 78, 91, 101, 102, 121, 127, 136, 163
Negentropy, 16
Neoclassical economics
  failures, 159
  natural resources, 109, 110
Newton, Isaac, 7–9, 21
Nutrients, 27, 81, 89

O
Odum, Eugene, 49, 161
Odum, Howard, 19, 75–79, 89
Oil, 11, 19, 22–27, 29, 30, 38, 39, 41–45, 53, 85, 90, 93, 95–99, 103, 107–116, 120–126, 131, 134, 136, 137, 139, 146–155, 159–161
Opportunity costs, 4, 64
Organic matter, 39, 41

P
Pareto efficiency, 101
Peak oil, 85, 146, 150, 155
Pennsylvania, 139, 161

Perlin, John, 89, 92
Petroleum, 23, 28, 41–43, 45, 95, 97, 111, 112, 114, 122–124, 130, 146, 152, 158, 159, 163
Phosphorus, 30
Prices, 41, 43, 93, 100, 101, 110, 111, 113, 124, 128, 151, 153, 160
Primary productivity, 81
Production, 7, 11, 19, 27, 41–44, 55–57, 66, 76, 81, 83, 84, 91–93, 95, 97, 98, 100, 102–104, 107–112, 114, 120, 122, 124, 129, 130, 137, 139, 140, 146–152, 159–162
  biological, 55, 56, 66
  economic, 27, 66, 83, 84, 92, 95, 97, 98, 100, 102, 104, 107–110, 124, 125, 150
  oil, 42, 43, 98, 103, 110, 146–152
Productivity
  economic, 83, 84, 110, 124, 129
  growth, 22, 45, 49, 55, 66, 67, 101–103, 145, 146, 149, 153, 155–162, 167

R
Rain shadow, 8, 37
Reserves, 45, 70, 110, 116
Resources, quality, 15, 41, 45, 102, 109–111, 127, 134, 163
Ricardo, David, 41, 109

S
Science
  biophysical, 101, 102, 146, 159–161, 167
  natural, 98, 100, 101, 159
  neoliberalism
  social, 100, 101, 151
Scientific method
  need for, 8, 9
Secular stagnation, 159, 162
Solar energy, 28, 37–39, 51, 91, 92, 95, 97, 136, 162
Solar radiation, 28, 162
Steel, 30, 92, 93, 123, 124, 125
Sustainability, 145, 146, 153, 155
Sustainable development, 92, 93, 145, 159, 162
Systems approach, 11–13, 16, 18, 42, 54, 74, 75, 79, 81–85, 115, 120, 129, 133, 134, 136–138, 151, 162

T
Tainter, Joseph, 89, 92, 145
Technology, resource quality, 41, 45, 109, 134, 163
Thermodynamics, 7, 10, 11, 13–15, 19, 24, 28, 35, 49, 50, 54, 73, 74, 85, 159, 162

Thompson, Benjamin (Count Rumford), 9
Tradeoffs, 64–65, 71
Transportation, 11, 22, 95, 122, 128, 155, 162
Tropics, 36, 81

U
Unemployment, 101
Uranium, 24, 39

W
War, 5, 22, 89
Watt, James, 97
Work, 4, 7, 9–17, 22–26, 28, 29, 35, 37, 52–55, 61, 62, 75–78, 80, 92, 95–97, 100, 109, 158, 166, 167

CPSIA information can be obtained
at www.ICGtesting.com
Printed in the USA
LVOW02*0330230817
546039LV00002B/2/P

9 783319 478203